Gufu Oba

Pastoralist Resilience to Environmental Collapse in East Africa since 1500

palgrave
macmillan

Gufu Oba
Norwegian University of Life Sciences
Ås, Norway

ISBN 978-3-031-48290-8 ISBN 978-3-031-48291-5 (eBook)
https://doi.org/10.1007/978-3-031-48291-5

This Palgrave Macmillan imprint is published by the registered company Springer Nature Switzerland AG.
The registered company address is: Gewerbestrasse 11, 6330 Cham, Switzerland

Paper in this product is recyclable.

Pastoralist Resilience to Environmental Collapse in East Africa since 1500

I dedicate this book to the Obbu Borana, whose knowledge of pastoralism in the past and the present has enriched this work.

Acknowledgments

This book has had a long gestation. Initial work started in 1983, and the writing work was completed 29 years later. The book's findings will provide a basis for training the local communities, where we aim to strengthen those indigenous adaptive strategies that sustain the resilience of pastoralism to allow the community to chart their future.

The author acknowledges funding from the Faculty of Landscape and Society (LANDSAM) at the Norwegian University of Life Sciences that supported the writing of this research. The Department of International Environment and Development Studies (NORAGRIC) funded part of the fieldwork from 2003 through 2011. Data collection from the 1980s through the 1990s was funded using various sources.

The work reported in this volume involved numerous individuals who participated in data collection and provided cultural and historical information from as early as 1983. I want to thank Chief Galma Dabasso, whose letter motivated this long-term research (Chap. 6). My long-term assistant, Bagaja Jirmo Sap'ansa, who had spent many months collecting and transcribing data, is acknowledged. The author also acknowledges others who contributed to field data collection: Waqo Gufu, Dido Daki, Kana Wako, and Halkano Malicha. Others who provided historical information on pastoral collapse are Dr. Borbor Bulle and Gayo Boru Aga. For historical memory, I cannot fail to thank the late Galma Godana Happi, Mwalim Bule Gufu, Qalicha Charfana, Jaldesa Boru Elema, Huqa

Morowa, and Dabassa Mudale[1] for contributing to this study. Appreciation is extended to the elders who served in the case studies, such as Guracha Dambala, Guyo Qae Dabasso, Rev. Shedo Oba, and Paul Guyo Bonaya, and numerous women who shared their knowledge about social capital networks. Others not mentioned here are found in the text of different chapters.

My long-term editor, Dr. Jill Fresen, paid close attention to the presentation's quality, increasing the work's clarity. Three anonymous reviewers are thanked for their thoughts and comments on the earlier manuscript. Dennis Milewa is thanked for drawing the maps. My final gratitude goes to the Obbu people who contributed to this book's making. The results of this work were presented to the elders from the region in 2008, 2011, and 2013. This work will be the basis for a Pastoralist community training institute. The institute's purpose will be to develop teaching and learning by the elders—men and women and the youth, on matters that affect their lives and to promote their involvement in finding local solutions to the problems documented in this book.

Ås, Norway
November 2022

Gufu Oba

[1] These elders have since passed on, but their story will go into the annals of history for the Obbu Borana.

Contents

About the Author

Gufu Oba is a professor emeritus in the Faculty of Landscape and Society (LANSAM) at the Norwegian University of Life Sciences. Oba's work is highly interdisciplinary, combining ecological sciences and social and environmental history. His previous books include *African Environmental Crisis* (Routledge, 2020), *Herder Warfare in East Africa* (White Horse, 2017), *Climate Change Adaptation in Africa* (Routledge, 2014), and *Nomads in the Shadows of Empires* (Brill, 2013).

List of Figures

List of Tables

CHAPTER 1

Managing Resilience in Pastoralism: An Introduction

The question of resilience or catastrophic collapse of pastoralism portrays binomial views on alternative outcomes in response to disasters. Primarily, conventional wisdom is that pastoralism, having been exposed to disasters over centuries,[1] would have developed mechanisms for adjusting to environmental catastrophes.[2] In a resilient pastoral system, at least hypothetically, livestock, human populations, the environment, and social security systems would never reach extinction (our approach in this work) but may be transformed. Contemporaneously, however, a picture emerges that pastoralism faces a crisis[3]—threatened with complete collapse.[4] Anatoly Khazanov,[5] perhaps referring to this persistent contemporary pastoral

[1] Smith, A. B., *Pastoralism in Africa: Emergence of pastoral traditions* (London, Hurst & Co., 1992).

[2] Oba, G. *Climate change adaptive strategies in Africa: an history ecology* (London, Routledge, 2014).

[3] Little, M. A., Dyson-Hudson, R., Leslie, P. W. and Dyson-Hudson, N., 'Framework and theory' in Little, M. A, and Leslie, P. W. (Eds) *Turkana herders of the dry savanna: Ecology and biobehaviuoral response to nomads to an uncertain environment* (Oxford: Oxford University Press, 1999).

[4] O'Connor, T. G. and Kiker, G. A., 'Collapse of the Mapangwe society: Vulnerability of pastoralism to increasing aridity,' *Climate Change* 66 (2004): 49–66.

[5] Khazanov, A. 'Pastoral nomads in the past, present and future: A comparative view,' in Olson, P. (Ed.) *The struggles for the land: Indigenous insight and industrial empire in semiarid world* (University of Nebraska Press, 1990).

G. Oba, *Pastoralist Resilience to Environmental Collapse in East Africa since 1500*, https://doi.org/10.1007/978-3-031-48291-5_1

crisis debate, states: 'Were I to put my ideas about pastoral nomads in a nutshell, I would express them as follows: Their past was unique, their present is transition…and as for the future—they have no future at all if present tendencies continue.' Johan Helland[6] further captures this view: 'The most pessimistic outlook asserts that pastoralism as a way of life…has outlived its well-established successes and that ecological crisis, famine, dependency [on state emergencies] and permanent destitution' would denote a collapsing system.

The views about the perils of pastoralism may seem antiquated given that recent scholarship has depicted pastoralism as a highly adaptive system of resource management contrary to the above sentiments.[7] The work presents strong evidence that pastoralism is not static; it has been adjusting to all the shocks and trying to operate in the risky and volatile environment in the Horn of Africa. The analysis of the historical context shows that the pastoral system and its customary institutions enabled the society to survive, but also shows the considerable pressure under which the system has been operating. This can be assessed from perspectives of impacts of colonial control, wars on changes in trajectories of land use, individual drought management strategies, courses of long-term changes in livestock demographics and household wealth, and responses by gendered social capital networks (This study).

The survival of pastoralism relies primarily on access to natural resources and the ability to maintain mobility in a highly variable environment.[8] More precisely, the existence of pastoralism is coupled with the presence of social-political systems that document memories of the past and present. Equally, due to external factors and internal changes, the loss of grazing lands would undermine pastoral resilience.[9] More fundamentally, the collapse of pastoralism and land use transformation—combined with

[6] Helland J., 'Participation and governance in the development of Borana: southern Ethiopia,' in Salih, M. A. (Eds). *African pastoralism: Conflict, institutions, and government* 2001.

[7] Behnke, Roy H., and Carol Kerven. *Redesigning for risk: tracking and buffering environmental variability in Africa's rangelands.* Vol. 1. (London: Overseas Development Institute, 1994); Behnke, Roy H., and Ian Scoones. Carol Kerven, eds. *Range ecology at disequilibrium: new models of natural variability and pastoral adaptation in African Savannas.* (London: Overseas Development Institute, 1993).

[8] Salzman, P.C. *Pastoralists: Equality, hierarchy, and the state* (London: Routledge, 2018).

[9] Galaty, J., 'The Maasai group ranch: politics and development in an African pastoral society,' in Salzman, P. C. (Ed) Galaty, John G., et al. "The future of pastoral peoples." *Conference on the Future of Pastoral Peoples, Nairobi (Kenya), 4–8 Aug 1980.* IDRC, 1981.

climatic shocks—would trigger livestock die-offs, inducing dynamic changes in household wealth. So, in a collapsed pastoral system, social capital networks that provide everyone with livelihood security—may be transformed. Primarily, the transformation would not return the system to the pre-crisis *ante* but would need to seek a different level of existence.[10] This is a persistent challenge to African pastoralists in general.

The research reported here describes pastoral resilience or catastrophic collapse over three historical periods. It begins with a pre-colonial regional investigation (1530–1900), using social memory to show cycles of stability or failure. During the colonial period (1908–1962), it examines how the trans-frontier control of herd grazing movements transformed adaptive strategies in response to administrative and climatic shocks. Following the independence of Kenya (1963–1980), the work evaluates how warfare broke pastoral resilience and changed land use. It further examines how the collapse of the pastoral economy motivated the cultivation of crops and the establishment of private pastures—transforming the adaptive drought management strategies of individual pastoral households. Also, during the post-independence period (1982–2011), the work assessed long-term livestock populations and household wealth dynamics in response to multiple droughts.[11] Additionally, it examines how pastoral collapse transformed clan social capital networks (1991–2012) into a broad-based network. Further, it evaluates how women's neighborhood food security networks were sustained under the transformed agropastoral production (1986–1996). The primary analysis is discussed within broader adaptive strategies to disasters.

The Main Arguments

The book is intended to provide a longitudinal overview of the varying responses of Borana Oromo pastoralists of southern Ethiopian and northern Kenya borderlands (Fig. 1.1) to changing climatic variability, systems of natural resources management and governance, socio-ecological conditions, external geopolitical events and shifting administrative regime

[10] Robinson, L. W. and Berkes, F., 'Applying resilience thinking questions of policy for pastoral systems: Lessons from the Gabra of Northern Kenya,' *Human Ecology* (2010) 38: 335–350.

[11] McPeak, J., Little, P. D. and Doss, C. R., *Risk and social change in an African rural economy: Livelihood in pastoral communities* (London: Routledge, 2012).

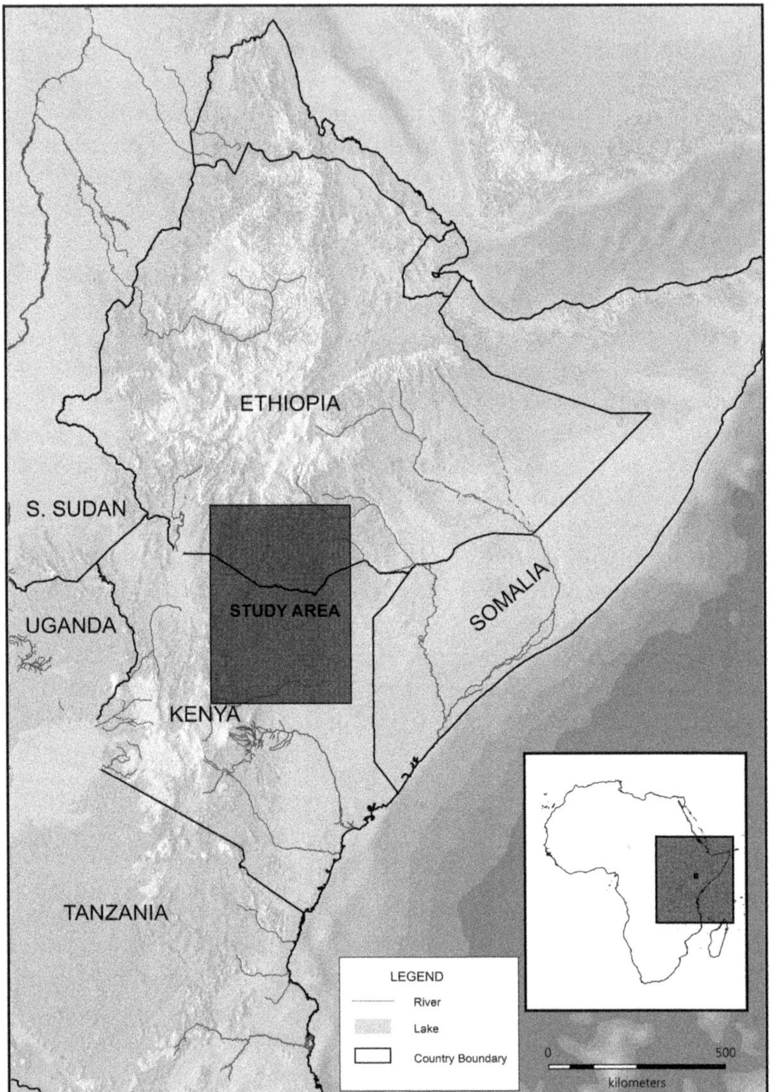

Fig. 1.1 Geographical location of the study area

interventions by the colonial nation-state and post-independence's laissez-faire policy responses. Internally, pastoralists used diverse adaptive strategies to seek new sources of livelihoods while culturally attempting to maintain institutions of pastoralism.

A distinctive feature of the book is its utilization of indigenous structures of social memory to extend the temporal range of the study much further back in time, from the sixteenth century to the present, untypical for studies focusing on contemporary issues alone. This has the potential to provide greater clarity on the antiquity of certain forms of resource governance and other adaptive strategies of the Borana Oromo. Linking this historical to longitudinal contemporary case studies within the same landscapes of social memory brings history into a modern-day debate on pastoralist responses to changing adaptative strategies. Indeed, the primary strength of this work is its engagement with a timeframe that is critically important for studies of pastoralism in exploring interdisciplinary studies on managing resilience.

In doing so, the work seeks to draw broader conclusions about the resilience of pastoralist systems and the changes that have occurred within adaptative strategies over long-term socio-economic, ecological, and political changes. The book emphasizes how disaster episodes, as critical historical events, are used to structure accounts of the past over centuries. Its central analysis is organized around the Borana *gada* timeline and corresponding event history calendars. In doing this, the work emphasizes the value of African pastoralist oral histories to contemporary interpretations of pastoralist resilience and historical interpretations of pastoralist livelihood histories more generally. The work carefully appreciates multiple perspectives and uses evidence-based analysis of pastoralists' disaster adaptations. In this sense, it is of fundamental value to an array of disciplines exploring the region's history in question and, more broadly, across northeast Africa. This is because the case studies are situated in the broader region that hosts linguistically numerous other pastoralist cultures, each with distinct histories, modes of historical analysis, livelihood systems, and socio-political organizations.[12] While the trends of change described in the current work would also apply to several other communities in the

[12] Schlee, Günther. *Identities on the Move: clanship and pastoralism in northern Kenya*. (London, Routledge, 2018); Spencer, Paul. *The Samburu: A study in geocentracy*. (London, Routledge, 2004); Spencer, Paul. *Nomads in alliance: symbiosis and growth among the Rendille and Samburu of Kenya*. (Oxford University Press, 2012).

geographical region, the author will resist extrapolating the findings to other cultural groups— the apparent reasons being different cultural and historical differences and types of pastoralism and ecological systems. Yet important lessons would indeed be achieved from the present work by researchers among other groups. The model has been used in recent pieces for the Lake Baringo and the Lake Turkana Basins, [13] which informed the current study.

This fact is of critical importance considering the massive negligence of the histories of pastoralist communities' indigenous knowledge of disaster survival and management.[14] Historical analyses that span several centuries are entirely missing in many cases. The author uses interdisciplinary historical research to address this deficiency by engaging with Borana Oromo's oral histories, which appear to be memorized by oral historians and time experts for over five centuries, unlike in other pastoralist communities in northern Kenya. Based on research spanning some three decades, the author engages with the complexity of history telling in the changing world of Borana pastoralism, considering the contemporary socio-political contexts of this trans-frontier region.

However, it is essential to inform the readers that this work is yet to focus on development interventions that are popular in the literature on contemporary pastoralism.[15] Instead, the aim is how this case community, on its own, responded to historical disasters and the changes experienced. The reasons why a frontier region is considered will be evident in the context of this historical study at social, political, and ecological scales. This work will attempt to use the findings from the historical and case studies and make concrete proposals on emerging adaptive strategies that the population might consider using in the future (see Chap. 10). These counter the oft-ambiguous assumptions among the outside world that

[13] Bollig, Michael, and David M. Anderson, eds. *Resilience and Collapse in African Savannahs: Causes and consequences of environmental change in East Africa* (London, Routledge, 2018); Samuel F. Derbyshire, *Remembering Turkana: Material Histories and Contemporary Livelihoods in North-Western Kenya,* (London, Routledge, 2021).

[14] Akyyeampong, E., Bates, R. H., Nunn, N. and Robinson, J. A. 'Introduction: Africa—the historical roots of its underdevelopment,' in Akyeampong, E., Bates, R. H., Nunn, N. and Robinson, J. A. (Eds.) *Africa's development in historical perspective* (New York: Cambridge University Press, 2014), 2; Beinart, W. and Coates, P., *Environment, and History: The taming of nature in the USA and South Africa* (London: Routledge, 1995).

[15] Oba, G., *African environmental crisis: A history of science and development* (London: Routledge, 2020).

pastoralists lack solutions to problems associated with disasters. This counter-narrative is the strength of this work.

The work focuses on the impacts of the diversity of disasters on pastoralism. The work contends that it is possible to capture the evolution of pastoralists' adaptive strategies to cycles of disasters by shifting analysis from general to specific case examples—following a community through time and understanding how event history influenced their adaptive strategies.

The work draws lessons from the *long-durée'* methods of the *Ecoles des Annales* school, which have been popular for discussions on historical, spatial, and social-political contexts.[16] The *Annales* school methods include, among others, focusing on the strong associations between the environment and societies and providing explanations for successive events that trigger change.[17] Consequently, where specific environmental events induce regional scale changes, a clear idea about the mechanisms behind changes in human adaptations might be uncovered.[18] According to Miller, social memory frames can be used to identify past and present relationships.[19]

Social memory is embedded in local institutions[20] and describes the practice of transmitting knowledge orally from one generation to another.[21] Social memory records historical and socio-economic changes and shifts in environmental conditions.[22] The methodology of gathering and documenting social memory allows regional and longitudinal local

[16] Ley Roy Ladurie, E., *Times of feast, times of famine: A history of climate since the year 1000* (New York: Garden City, 1971); Watts, M., *Silent Violence: Food, famine, and peasantry in northern Nigeria* (Berkeley: University of California Press, 1983).

[17] Stoianovich, T., 'French historical method: The Annales paradigm,' in Stoianovich, T., (Ed.) *French historical method: The Annales paradigm* (Cornell University Press, 1976), 19.

[18] Jansen, J., 'Habitation and warfare strategies in nineteenth century Mande—a view from the Kafu,' in Howard, A. M and Shain, R. M. (Eds.) *The spatial factor in African history: The relationship of the social, material, and perceptual* (Leiden: Brill, 2005).

[19] Salzman, P. C., 'Introduction,' in Shutlive, V. H., Altshuler, N. and Zamora, M. D., (Eds.) *Contemporary nomadic and pastoral peoples: Africa and Latin America* (Williamsburg: Virginia, USA, 1982), ix.

[20] Lamb, H. H., *Climate, history and the modern world* (London: Routledge, 1982).

[21] Crumley, C.L., 'From garden to globe: Linking time and space with meaning and memory,' in McIntosh et al. (Eds.) *The way the wind blows: climate, history, and human action* (Columbia University Press, 2000).

[22] Fagan, Brian. *The great warming: Climate change and the Rise and fall of Civilizations.* Bloomsbury Publishing USA, 2008, 81.

analysis of how disasters influenced pastoralists' levels of resilience or induced a collapse in the past and present.[23] While absolute dating might be difficult, social memory helps reconstruct a traditional event history calendar (EHC).[24] If applied to climate variability in a historical context, social memory is not only a source of knowledge but also provides practical past experiences transmitted to the present generation. Social memory calendars were then compared with proxy environmental indicators to predict the returns of disaster cycles.[25] By investigating society's adaptive strategies under changing socio-political, economic, and ecological conditions,[26] the work illustrates the levels to which pastoralism has been resilient, collapsed, or transformed in response to multiple disasters. We will next conceptualize resilience and collapse to analyze socio-ecological systems in response to disasters.

Framing Resilience and Collapse

Before proceeding, it is essential to tightly define such terms as 'adaptation, resilience, collapse, and transformation' as used in this work. We will also suggest the metrics by which each concept will be measured. This would involve selections of substantive indicators of resilience or collapse instead of 'transformation.'

Adaptation Strategies

Adaptive management strategies are designed to maintain resilience in the system—which for African herders is scale dependent; that is, they function at local and geographical scales.[27] Adaptation might also vary at family

[23] Beinart, W. and Coates, P., *Environment, and History: The taming of nature in the USA and South Africa* (London: Routledge, 1995).

[24] Henige, David Patrick. *The chronology of oral tradition* (Oxford: Clarendon Press, 1974).

[25] Oba, G., *Climate change adaptation in Africa: a historical ecology* (London: Routledge, 2014).

[26] Randall, S., 'Where have all the nomads gone? Fifty years of statistical and demographic invisibilities of African mobile pastoralists,' *Pastoralism: Research, Policy and Practice* (2015): 5: 1–22, https://doi.org/10.1.1186/s13570-015-0042-9.

[27] McIntosh, R. J., Tainter, J. A. and McIntosh, K. 'Climate, history and human action' in McIntosh, R. J., Tainter, J. A. and McIntosh, K (Eds.) *The way the wind blows: Climate, history, and human action* (New York: Colombia University Press,

and community levels and from one period of time to another.[28] Adaptive strategies offer choices between various courses of action.[29] Adaptation depicts responses to uncertain predicaments that influence subsequent decisions in the context of global and regional climatic variability.[30] Decisions are based on established knowledge, and where successful, adaptations would depend on changes in social behavior in response to disasters. The ability to make decisions supposes social stability, availability of labor, local and geographical resources, efficient methods of information gathering, family herd sizes, and management options.

Accordingly, in a drought-prone region, individual decisions would be contingent on the availability of grazing and water resources.[31] Where adaptation is used to mitigate environmentally induced crises, the choices are on what is practical, with each option presenting different degrees of vulnerabilities.[32] Transformation is, therefore, a process that can result in different outcomes and requires individuals and societies to plan accordingly.[33] For herders, the goal is to avoid those strategies that might undermine survival and promote those that enhance them.[34] Consequently, mitigation mechanisms might change in response to shifts in

[28] Smith, A. B. *African herders: Emergence of pastoral traditions* (Lanham USA: Altamira, Walnut Creek, 2005).

[29] Orlove B. 2009, 'The past, the present and some possible futures of adaption,' in Adger, W. Neil, Suraje Dessai, Marisa Goulden, Mike Hulme, Irene Lorenzoni, Donald R. Nelson, Lars Otto Naess, Johanna Wolf, and Anita Wreford. "Are there social limits to adaptation to climate change?" *Climatic change* 93 (3) (2009): 335–354.

[30] Handmer, J. W., Dovers, S. and Downing, T. E., 'Societal vulnerability to climate change and variability, mitigation, and adaptation,' *Strategies for Global Change* 4 (1999): 267–281; Zampalingre, N., Dossa, L. H. and Schlecht, E., 'Climate change and variability: Perception and adaptation strategies of pastoralists and agro-pastoralists across different zones of Burkina Faso,' *Reg. Environmental Change* 14 (2014): 769–783.

[31] Smith, A. *Pastoralism in Africa: Origin and development ecology* (London: Hurst & Co., 1992, 2, 127).

[32] Nyong, A., Adesina, F. and Elesha, B. O., 'The value of Indigenous knowledge in climate change mitigation and adaptation strategies of African Sahel,' *Mitigation, Adaptation Strategies of Global Change* 12 (2007): 787–797.

[33] Toulmin, C., 'Tracking through drought: options for destocking and restocking' in Scoones, I. (Eds.) *Living with uncertainty: new directions in pastoral development in Africa.* (London: ITP, 1994).

[34] Smit, B. and Wandel, J. 'Adaptation, adaptive capacity and vulnerability,' *Global Environmental Change* 16 (2006): 282–292.

environmental conditions.[35] When adaptation fails, collapse triggers economic transformation.[36] Depending on the perturbation pressure and frequencies, the triggers might collapse systems—transforming pastoralism into agropastoralism. The process hastens the resettlements of former pastoralists, resulting in the abandonment of mobile land use and seeking survival in peri-urban and urban environments using non-pastoral livelihood strategies. In the changed system, individual households could diversify their livelihoods and construct neighborhood social capital networks to supply social insurance for the members.

Resilience and Catastrophic Collapse

How the term 'resilience' is understood in the current work must be highlighted at the onset. Contrary to the mainstream literature that used the concept in the context of equilibrium ecological systems, what is being described is a disequilibrium system. This means that the way it is used here explores complex analytical methods in both space and time that dispels the plans from being predictable in the sense of cyclic changes described in a Social-Ecological Systems (SES) framework to triggers of disasters that reflect resilience, collapse, and transformation. The changes, contrary to a general perception, are unpredictable. Nonetheless, one would still apply the cyclic model about the regenerative phase (r), followed by accumulation and maturity (k)[37] in range and pastoral productions. The changes in such a complex system might be interlinked and exist discretely. Thus, applying the dynamic model oversimplifies an overly complex system. How would one, for example, relate a collapse in stock demography to shifts in household wealth? The two approaches are interlinked, but their outcomes are expressed differently. Yet, essential

[35] Agarwal, S. K., 'Emerging technological intervention models with scalable solutions for adaptions to climate change and livelihood gains in Indian Himalayan region: Case studies on action research at the grassroots level,' in Nantiyal, S. et al. (Eds), *knowledge systems of societies for adaptation and mitigation* (Berlin: Springer-Verlag, 2013).

[36] Middleton, G. D., 'The show must go on: Collapse, resilience, and transformation in 21st -century archeology,' *Reviews in Anthropology* 46. 2–3 (2017): 78–105; Abel, N., Cumming, D. H. M. and Anderies, J. M., 'Collapse and reorganization in social-ecological systems: Questions, some ideas, and policy implications,' Ecology and Society 11.1 (2006): 17 [online] URL: http://www.ecologyandsociety.org/vol11/iss1/art 17/.

[37] Holling, C. S., 'Resilience and stability of ecological systems,' *Annual Review of Ecology and Systematics* 4 (1973): 1–23.

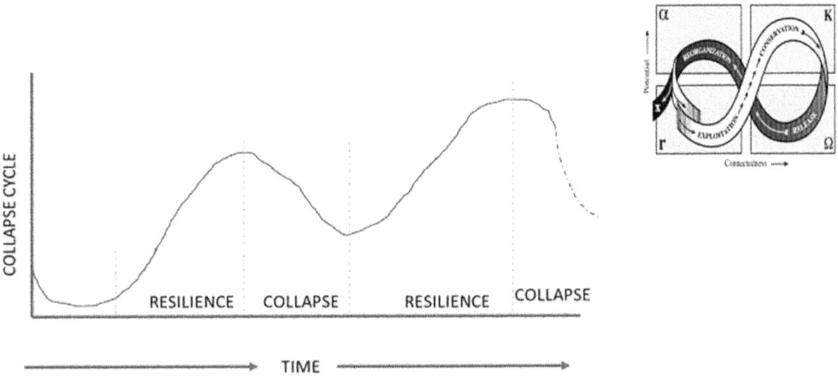

Fig. 1.2 Disaster cycles: The cycles might be compared with the phases of an adaptive cycle [Inset] *r*: regenerative phase, *k*: maturity phase, *Ω*: release phase (collapse), and *α*: reorganization

conclusions would be made regarding systems' behavior toward intensifying disaster processes. Depending on the frequencies and intensification of disasters, a collapse (Ω) is followed by reorganization (α) (see inset, Fig. 1.2). The latter process might change the system's short- and long-term trajectory. This happens during multi-year disasters. The changes might be patchy and discontinuous on both spatial and temporal scales.[38] From historical analysis (see Chap. 3), the changes in response to decadal disasters might overwhelm traditional adaptive systems when the system is described to have collapsed. The regenerative potential and the time involved will determine whether the system recovers and restores resilience or shifts to a transformed state.

Consequently, 'resilience' is a term that has come to be deployed in analytically diverse ways across anthropology, history, archaeology, ecology, and development studies.[39] For this reason, this work would grapple with the complex ways the term is used. More specifically, in

[38] Redman, C. L. and Kinzig, A., 'Resilience of past landscapes: Resilience theory, society, and the *longue dure'e,'* Conservation Ecology 7.1: 14 [online] URL: http://www.consecol.org/vol17/iss1/art14/.

[39] Taşan-Kok, Tuna, Dominic Stead, and Peiwen Lu. 'Conceptual overview of resilience: History and context,' *Resilience thinking in urban planning* (2013): 39–51; Vernon, R. Fox. 'A brief history of resilience,' *Community planning to foster resilience in children*. Springer, Boston, MA, 2004. 13–26; Osterhoudt, Sarah. 'Remembered resilience: oral history narratives and community resilience in agroforestry systems,' *Renewable Agriculture and Food Systems* 33.3 (2018): 252–255.

disequilibrium arid and semi-arid ecosystems, ecological functioning is controlled by factors such as rainfall, droughts, and other natural disasters.[40] The events influence decisions by individual herders to practice novel management methods by moving their stock and applying personal knowledge to adjust their strategies to stresses induced by disaster events.[41] Alternative thinking also exists if factors such as collapse and recovery in stock populations or collapse and recovery in household wealth are considered. The goal is to maintain systems resilience,[42] which might be interpreted differently. In both, the matrices for measurements are a stock population to human capital and variable rainfall.

Resilience describes how climate variability influences herd movements in addition to political controls that play an opposite role. The total outcome describes how and why resilience is sustained, or catastrophic collapse is anticipated. Thus, the work defines pastoral resilience as the ability of the pastoral system to respond to disasters to reduce vulnerabilities and manage transformation. Then, the next challenge is how to measure a transformed system in the context of resilience or collapse. The outcomes of transformation might be in terms of changes in land use, such as the adoption of agropastoralism and semi-privatization of the communal grazing lands that would influence how the communities respond to the changes.

Droughts—the absence of rainfall when expected collapses farming and animal husbandry. The collapse in crop cultivation occurs in the short term, while the failure in animal husbandry is a long process before recovery is accomplished. The decisions by the farmer/herder would, therefore, depend on whether to abandon one system and shift attention to the other. The response in both cases might vary from the changes experienced by the ecosystem. The functioning of the ecosystem would be influenced by climate variability, but the decisions of the farmer/herder affect the way the disaster impacts are managed. For this reason, greater emphasis is placed on how the system recovers or transforms over prolonged

[40] Holling, C. S., 'Resilience and stability of ecological systems' *Annual Review of Ecology and Systematics* 4 (1973): 1–23.

[41] Brunson, M. W., 'The elusive promise of social-ecological approaches to rangeland management,' *Rangeland Ecological Management* 66 (2012): 632–637.

[42] Baival, B., *Community-based rangeland management and the social-ecological resilience of rural Mongolian communities, Doctor of Philosophy, Department of Forest and Rangeland Stewardship, Colorado State University* (2012).

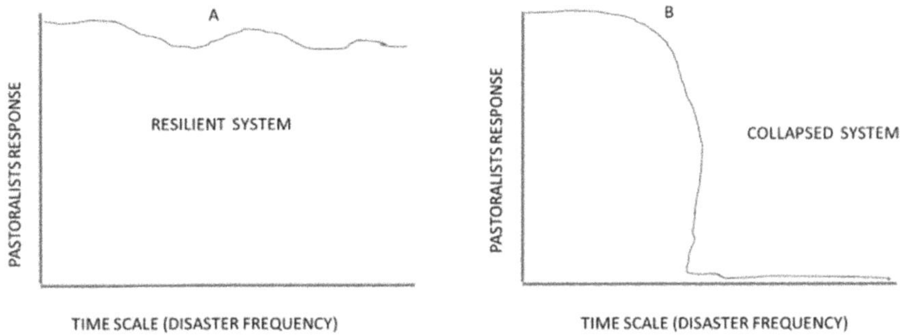

Fig. 1.3 Time scale disaster frequencies and pastoral responses in (A) resilient and (B) collapsed systems

periods after disaster episodes.[43] When combined, the persistence of ecological, social, political, and economic systems would promote sustained resilience (Fig. 1.3A). However, disasters may act independently on different resilience structures and induce a total collapse (Fig. 1.3B).

According to Jared Diamond,[44] processes leading to collapse may be either sudden or gradual. Sometimes, failure is traumatic—caused by droughts, famine, epidemics, and warfare. In gradual changes, subsistence production is undermined when environmental parameters do not support production or ecological and economic crises lead to acute stress, which triggers a collapse.[45] In the described system, resilience (recovery) and failure spontaneously operate. All plans would transform, such as former herders adopting crop cultivation and clan social capital networks changing to broader systems of social networks. In contrast, resilience in women's social capital network might include utilizing formal and informal methods—the former relying on external resources to support the latter networks.

[43] Gitz, V. and Meybeck, A., 'Risks, vulnerabilities and resilience in a context of climate change,' in Meybeck, A., Lankoski, J., Redfern, S., Azzu, N. and Gitz, V. (eds) *Building resilience for adaptation to climate change in the agricultural sector*, Proceedings of a Joint FAO/OECD Workshop, Rome, 23–24 April 2012.

[44] Diamond, J., Collapse*: how societies choose to fail or survive* (New York and London: Viking Penguin/Allen Lane, 2005).

[45] Butzer, K. W., 'Collapse, environment, society' *PNAS* 109 (10) (2012): 3632–3639; O'Connor, Tim G., and Gregory Alan Kiker. "Collapse of the Mapungubwe society: vulnerability of pastoralism to increasing aridity." *Climatic Change* 66.1 (2004): 49–66.

Almost invariably, therefore, multiple collapses have occurred throughout the centuries. However, those widely reported happened in the late nineteenth century when the combined effects of epidemics, famine, warfare, and ecology altered the course of history.[46] During the crises, the collapse of livestock populations was followed by the failure of human demography.[47] Under the circumstances, social and political structures disintegrated, and survivors dispersed among hunting and gathering communities. Social collapse is a large-scale transformation of social institutions and social capital networks. Depending on the extent of change, the shift may preserve vestiges of traditional institutions that assume new functions. On the other hand, if there have been high frequencies of perturbations, livestock populations may not reach the pre-collapse threshold following recovery. They would, therefore, remain vulnerable until the next disaster.[48] Increased frequency of disasters would cause pastoral production to decline until the system is transformed. In the changed scenario, new adaptive strategies might be advanced. In such a system, the management might not focus on restoring the historical resilience benchmarks partly because the social, political, and economic technologies and knowledge for rejuvenating the system do not exist. In the case of pastoralism, one would need to elaborate on the scales of changes that might be targeted for disaster management. We might simplify the analysis by focusing only on a single adaptive strategy that pastoralists use among the many simultaneously operated approaches.

Phases of Pastoral Collapse

An example will suffice here using pastoral collapse phases. In Camilla Toulmin's[49] three phases of pastoralists' responses to drought stress, the

[46] Oba, G. *Herder warfare in East Africa: A social and spatial history* (Winwick, White Horse Press, 2017).

[47] Tiki, Waktole, and Gufu Oba. "Ciinna–the Borana Oromo narration of the 1890s great rinderpest epizootic in Northeastern Africa." *Journal of Eastern African Studies* 3.3 (2009): 479–508.

[48] Dong, S., Liu, S. and Wen, L., 'Vulnerability and resilience of human natural systems of pastoralism worldwide,' in Dong, S., Liu, S., and Wen L. (Eds) *Building resilience of human-natural systems of pastoralism in the developing world: Interdisciplinary perspectives*, (Springer International Publishing Switzerland, 2016).

[49] Toulmin, Camilla. *Livestock losses and post-drought rehabilitation in sub-Saharan Africa* (1985).

first phase is characterized by a shortfall in pasture and water. Drought stress becomes widespread, the movement of herds is constrained, and herd numbers begin to fall due to livestock mortalities. In phase two, the focus is aggravated by further shortages of water and pastures, and drought escape areas become exhausted. Phase three might be broken by rainfall. Consequently, indigenous adaptive strategies might shift from average to wet-wet years to wet-dry or dry-dry year cycles.[50] In SES that exhibit disequilibrium—where the performance of pastoral production is directly influenced by variable rainfall[51]—production is optimal during the wet-wet year, end-to-end cycles. Under these conditions, water and pastures are plentiful, and progressive recovery from past disasters might be anticipated. Conversely, multiple droughts resulted in herd collapse, threatening families with long-term poverty.[52] After a dozen years of negative herd growth, families will naturally be forced to abandon pastoralism.[53] The implication is that a change in one aspect of the system triggers changes in other systems—in our case, by relating livestock demographic changes to changes in the wealth status of families—the changes illicit varieties of decision-making processes by individuals. One of the consequences is changes in household decision-making in response to changes in household wealth to which we turn now.

Household Decision-making

For those left with stock, drought management adaptive strategies invariably required planning, setting priorities, and evaluating actions.[54] The geographical distribution of resources, access rights, and labor requirements influenced decisions. The loss of livestock during periods of severe droughts shifted household wealth. Before and after the collapse, the same

[50] Ayana, E. K., Ceccato, P., Fisher, J. R. B. and DeFries, R., 'Examining the relationship between environmental factors in pastoral areas of East Africa,' *Science of the total Environment* 557–558 (2016): 601–611.

[51] Boone, R. B. and Wang, G., 'Cattle dynamics in African grazing systems under variable climates,' *Journal of Arid Environments* 70 (2007): 495–513.

[52] O'Connor, T. G. and Kiker, G. A., 'Collapse of the Mapungwe society: Vulnerability of pastoralism to increasing aridity,' *Climate Change* 66 (2004): 49–66.

[53] Heffernan, C. "Livestock, Destitution and Drought: The impact of restocking on food security post-disaster." *Unpublished* (1998).

[54] Downing T. E., Ringius, L., Hulme, M. and Waughray, D., 'Adapting to climate change in Africa,' *Mitigation and Adaptation Strategies for Global Change* 2 (1997): 19–44.

households would go through transient wealth ranks where those wealthy once might become stockless under multiple droughts.[55] Understanding wealth shifts among families is crucial to explaining the loss of pastoral resilience under multiple droughts, which directly affected the functioning of social capital networks. With this background, the work briefly examines the contents of individual studies in terms of matrices for measuring resilience, collapse, and transformation.

The Matrices for Measuring Collapse, Resilience, and Transformation

For the regional analysis of the impacts of disasters on pastoralism from historical perspectives, we interviewed oral historians and time experts to reconstruct a traditional event history calendar (EHC) covering five centuries (Chap. 2). The social memory was then compared with proxy environmental indicators to predict returns of disaster cycles (Chap. 3). The methods for building social memory include oral poetry, folklore, and the geographical spaces surrounding historical events. Socio-political institutions also serve as a repository for social memory and are essential in passing knowledge from one generation to another. Using the Borana case, we evaluated how they used EHC. The survival of the Borana people is guided by strong institutions—such as the *gada*—which managed the stages of consolidation and the restoration of the pastoral economy, thus reorganizing the society and resource allocations to repair the damage that had been suffered.

At the turn of the twentieth century, colonial governments created frontiers that split nomadic societies under different colonial administrations (Chap. 4). The wet season grazing resources were left on the British side, while the dry season rangelands and water sources were left on the Ethiopian side of the frontier. Consequently, a trans-frontier grazing treaty became urgent. However, the treaty's implementation was subject to contentious disputes between frontier empires—first between Ethiopia and the British administration and later (1935–1940) between the Italian and British administrations.[56] From a detailed analysis of the impacts of climate

[55] Mwamfupe, D. G., and Claude G. Mung'ong'o. "Poverty and changing livelihoods of migrant Maasai pastoralists in Morogoro and Kilosa districts." (2003).

[56] Oba, G., *Nomads in the shadows of empires: contests, conflicts, and legacies on the southern Ethiopian-northern Kenyan frontier*. Brill, 2013.

history on pastoralists' adaptive strategies, we show five decades punctuated by recurrent droughts and heavy rainfall years returning in 10–15-year cycles. Pastoralism maintained resilience over five decades of colonial administration despite tremendous suffering and the livestock population collapsing.

This was until secessionist warfare had collapsed pastoralism by Kenya's independence (1963). The effects of the war were direct and indirect. Directly, the war crushed pastoralism, displaced populations, and impoverished families. Indirectly, the war triggered the transformation of land use into agropastoralism. The changes had significant impacts on the adaptive strategies of pastoralists in motivating crop cultivation and the development of private rangeland enclosures for preserving pasture, which continues to drive land values, as well as influence individual stock drought management strategies (Chap. 5).

Using biographic histories of ten herders, we examined individual drought management strategies from the 1980s to 2011 (Chap. 6). Drought management strategies varied among large and small herding units. Large herding units often used regional adaptive strategies, while small herding units practiced local and intensive drought management strategies (until the 2000s). During the 1990s, long-distance herd mobility across the southern Ethiopian–northern Kenyan frontier had ceased due to conflicts. Internally, due to the shrinking grazing lands, traditional herd splitting was abandoned by the 2000s. The changing adaptive drought management strategies show how, over three decades, the Obbu Borana pastoralists shifted their adaptations from regional to local.

Droughts' immediate and long-term impacts on stock population dynamics and household wealth[57] (Chap. 7) show how stock loss shifted individual households' wealth, moving from rich or medium to poor categories.[58] In 6 out of 29 years, cattle herds collapsed, thus altering herd structure in terms of the numbers of reproductive females, calves, and immature and mature males. The rates of herd recovery depended on time intervals of the returns of droughts. In 6 out of 12 years, rainfall was below average. The reproductive females disclosed more elasticity between herd collapse and recovery, more so than calves and immatures, while the

[57] Lesorogol, Carolyn K. *Contesting the commons: Privatizing pastoral lands in Kenya*. University of Michigan Press, 2008.

[58] Mwamfupe, D. G., and Claude G. Mung'ong'o. "Poverty and changing livelihoods of migrant Maasai pastoralists in Morogoro and Kilosa districts." (2003).

populations of mature males consistently remained low—due to offtakes. During the same period, household wealth shifted from higher to lower ranks. Droughts reduced household wealth, regardless of pre-drought class. The most volatile wealth ranks were the wealthy, the very rich, and the medium households; these categories of wealth ranks disappeared during droughts. About 24 percent of the households became stockless—meaning they were at risk of dropping out of pastoralism.

Further, the work reports on the impacts of droughts on the functioning of social capital networks (Chap. 8) at three levels. At the individual level, families build networks,[59] and at a mesoscale, community-wide social institutions[60] administer and operate social capital networks used by clans. At geographical scales, social capital networks are embedded in 'social capital network zones' across northern Kenya and southern Ethiopia frontier. Across all scales, the system shares the same rules and customs that bind each member.[61] The rules determine who has the right of access to stock and what criteria are used for herd sharing.[62] Depending on whether the system works at individual or mesoscale, transfers of stock from givers to receivers are negotiated. The aim is to guarantee poor members' survival as a form of insurance.[63]

The collapse in the pastoral economy created conditions that weakened clan social capital networks. Since traditional methods of stock distribution were no longer workable, the community from the 1990s adjusted to using neighborhood social capital networks that cut across different clans,

[59] Hardin, R., 'Gaming trust,' in Ostrom and Walker (Eds.) *Trust and reciprocity: Interdisciplinary lessons for experimental research*. Russell Sage Foundation, 2003; Pretty, Jules, and Hugh Ward. "Social capital and the environment." *World development* 29.2 (2001): 209–227.

[60] Laužikas, M and Dailydaite, 'Impacts of social capital on transformation from efficiency to innovation-driven business,' *Journal of Business Economics and Management* 16.1 (2015): 37–51; Schweizer, T. and White, D. R., 'Revitalizing the study of kinship and exchange with network approaches,' in Schweizer, T. and White, D. R. (Eds.) kinship, networks, and exchange (Cambridge University Press, 1998).

[61] Shivakumar, S. J., 'The place of indigenous institutions in constitutional order,' *Constitutional Political Economy* 14.3 (2003): 3–21. 2003.

[62] Lin, Nan. "A network theory of social capital." *The handbook of social capital* 50.1 (2008): 69, 12.

[63] Ostrom, E., 'Toward a behavioral theory linking trust, reciprocity, and reputation,' in Ostrom and Walker (Eds.) *Trust and reciprocity: Interdisciplinary lessons for experimental research*. Russell Sage Foundation, 2003.

thus changing the rules of resource sharing. Success, however, depends on how much it has helped members with wide-ranging needs.[64]

Further, social capital networks exclusively managed by women (Chap. 9) form part of the moral economy, encompassing the sharing of minute quantities of food (salt, tea leaves, sugar, and food in general) and other household services. The participants share social relations or neighbors where such relationships are constructed through the exchange of food and services.[65] Rural and peri-urban households were compared. In addition to informal networks, women have built women's groups to receive help from external sources.[66] The types of welfare enabled women to increase their negotiating power to access extra resources that, in turn, improved their families' livelihoods.[67] An important adaptive strategy of rural-peri-urban women's livelihood diversification is using credit and credit-clientship systems, relations between food sharing and household wealth, and the monetary values of food asked for. Finally, the work presents critical conclusions from the preceding analysis (Chap. 10) and makes proposals for managing emerging innovative mitigation strategies appropriate for future actions by the community.

[64] Forest, R. and Kearns, A., 'Social cohesion, social capital and the neighborhood,' *Urban Studies* 38.12 (2125–2143.

[65] Khalif, Zeinabu Kabale, and Oba, G., "'Gaafa dhaabaa-the period of stop': Narrating impacts of shifta insurgency on the pastoral economy in northern Kenya, c. 1963 to 2007." *Pastoralism: Research, Policy and Practice* 3.1 (2013): 1–20.

[66] Khalif, Zeinabu Kabale, and Oba, G., 'Continuity and Change in Social Networks among Borana Women.'

[67] Hobson, B. and Lindholm, M., 'Collective identities, women's power resources, and the making of welfare states,' *Theory and Society* 26 (1997): 475–508.

PART I

Regional Memories of Disasters

CHAPTER 2

Historical Research Methods: Regional and Local Cases

In this chapter, we seek simple methods to assess adaptive responses to disaster shocks and to understand the importance of spatial and temporal ecological and social variations. Through a 'bioregional lens,' new insights into understanding regional and local cases may shed light on the dynamics of adaptive strategies.[1] We used social memory to reconstruct traditional event history calendars (EHCs) to evaluate pastoralism's performance through multiple adaptation pathways. The social memory data, combined with other empirical data, was then used to develop a schematic framework illustrating how past socio-ecological events influenced human adaptations.[2] For historians whose research relies on evidence, the challenge remains how various environmental and social variables might be integrated into event history calendars.[3] Societies might explain these changes by focusing on the physical environment, local economies, social harmony/disharmony, and disaster hardships. Additionally, they might relate lifestyle changes with the return of disasters associated with social

[1] Mortzkin, G., Patterson, W. A. and Foster, D. R., 'A historical perspective on pitch pine-scrub oak communities in the Connecticut valley of Massachusetts,' *Ecosystems* 2 (1998): 255–273.

[2] Martyn, K. K. and Belli, R. F., 'Retrospective data collection using event history calendars,' *Nursing Research* 51.4 (2002): 270–274.

[3] Rabb, T. K., 'The historian and climatologist,' in Rotberg, R. I and Rabb, T. (Eds.) *Climate and history: Studies in interdisciplinary history* (Princeton University Press, 1981).

G. Oba, *Pastoralist Resilience to Environmental Collapse in East Africa since 1500*, https://doi.org/10.1007/978-3-031-48291-5_2

and political events.[4] In interdisciplinary methods, we evaluate pastoralism's performance through multiple adaptation pathways. The findings will be used to understand managing resilience in agropastoralism in the future. We present one example of how an event history calendar (EHC), combined with historical documentation, provides time-related trajectories of human adaptive responses in southern Ethiopia and northern Kenya (Fig. 2.1). While the scheme displays linear relations between time and actual event histories, the actual cyclic events (marked by wet, drought, and war years, disease episodic) designate different experiences during contemporary and historical periods.[5] We can use an example here. In the present study, we will use chronological event histories of the responses to disasters by the Borana pastoralists for five centuries. During the pre-colonial period, the Borana pastoralists suffered catalogs of catastrophes, resulting in collapses and a series of recoveries of pastoralism. The rinderpest pandemic of the late 1890s triggered the long-lasting memory of the failure. Before complete recovery was accomplished, the pastoralist's adaptive strategies were disrupted by the colonial partitioning of the region by the Ethiopian Empire and the Kenyan British Protectorate.[6] The colonial administrative control was superimposed on environmental variability. The creation of political frontiers did not alter pastoralism—while after the independence of Kenya in 1963, societal actions were driven by overarching governability imperatives and internal responses to external events. After independence, political agitation by the secessionist war supported by the Somali Republic crushed pastoralism. The collapse of pastoralism was followed by reorganization and transformation of land use. The changing circumstances—from pre-colonial, colonial, to post-independence times—provide glimpses of persistent disasters that transformed the adaptive strategies of the pastoralists in our case society.

One of the most exciting aspects of this study is its multiple dimensions of adaptive strategies in terms of space and time and the intertwining of political and social contexts. Using qualitative analyses, the work identifies different processes related to each other, directly or indirectly. Accordingly,

[4] Miller, J. C., 'The significance of drought, disease, and famine in the agriculturally marginal zones of west-central Africa,' *Journal of African History* 23 (1982): 17–61.

[5] Belli, Robert F., William L. Shay, and Frank P. Stafford. "Event history calendars and question list surveys: A direct comparison of interviewing methods." *Public opinion quarterly* 65.1 (2001): 45–74.

[6] Later, from 1935–1940, the Italians occupied and administered it before they were removed during the Second World War.

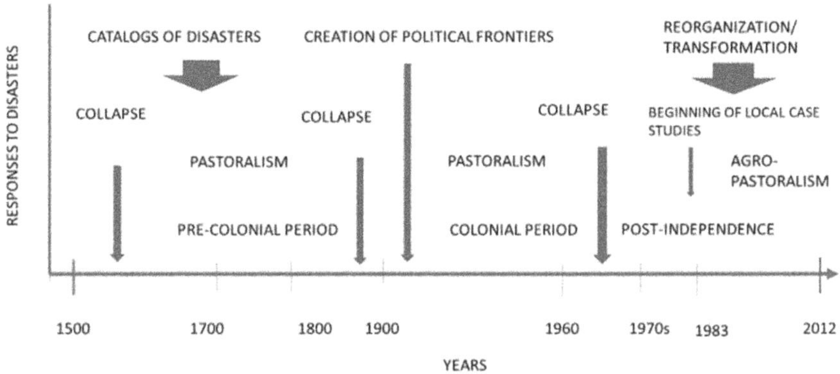

Fig. 2.1 A historical scheme showing dynamics of pastoralists' adaptive strategies in southern-central Ethiopia and northern Kenya between 1500 and 2012

referring to Fig. 2.1 again, we may summarize methods of resilience and the collapse of pastoralism set centuries ago. The triggers of stability or collapse varied over time, and the outcomes might have been different, depending on the nature of the stimuli. Concentrating on our analysis during the pre-and post-independence periods, one would realize that the time scale allows for identifying the trajectories of pastoralism and its transformation to agropastoralism. The latter, focusing on local case studies, provides a baseline that will be useful for managing resilience in agropastoralism in the future. We articulate these adaptations using socio-ecological and political-ecological methods.

Socio-ecological and Political-ecological Methods

Socio-ecological and political-ecological methods capture cross-scale and long-term responses to environmental disturbances. The political ecology describes the roles played by various actors using socio-ecological methods.[7] This raises important questions as to why researchers should focus on the power of social memory in recalling past events.[8] The triggers

[7] Bebjaminsen T. A. and Svarstad, H., Political ecology: A critical engagement with global environmental issues, Palgrave Macmillan, 2021.

[8] Rey Craft, R. M. and Bawden, G., 'Environmental disaster and the archaeology of human response: Introduction,' in Bawden, G. and Rey Craft, R. M. (Eds.) *Environmental disaster and the archeology of human response*, Maxwell Museum of Anthropology, Anthropological Papers # 7, Albuquerque, New Mexico, 2000.

represent historical processes as opposed to one-time events.[9] This approach is crucial for understanding how pastoralist adaptive behaviors changed across space and time. For African pastoralists, environmental, social, and political changes are typically explained in terms of the rainy season, with intervening dry spells determining changes in range production and herd structure. Responding to the patchy and erratic rainfall and temporally and spatially distributed pastures, herders historically used mobility as the most efficient means of managing climate variability.

Herders using systems memory capture details of past collapses and distant recoveries, as gauged in the present. This might be compared with the viewpoints of Charles Redman and Ann Kinzig[10] on resilience changes about historical shifts in rangeland production. In the first case, the changes are episodic, characterized by the slow accumulation of resources (i.e., herd growth and range production) that collapse in response to a series of ecological shocks. The second view describes processes as 'patchy' and 'discontinuous.' The third view is that changes involve multiple equilibria—involving various tipping points in cycles of resilience and collapses. The fourth view is that fixed land-use policies—ignoring numerous scales and temporal and spatial climatic variability—would undermine stability, triggering collapse processes. The collapse might be followed by recovery; alternatively, the system is reorganized and transformed.[11] This means the system is unstable, fluctuating from one period to another.[12] For African herders, mitigating the impact of disasters might require adopting aversive adaptive strategies, thus disclosing patterns of change in social behavior.[13] This is the case when pastoralism is transformed into agropastoralism.

[9] Collins, S. L., Carpenter, S. R., Swington, S. M., Orenstein, D. E. et al., 'An integrated conceptual framework for long-term social-ecological research,' *Frontiers of Ecological and Environment* 9.6 (2011): 351–357.

[10] Redman, C. L. and Kinzig, A., 'Resilience of past landscapes: Resilience theory, society, and the *longue dure'e,' Conservation Ecology* 7.1: 14 [online] URL: http://www.consecol.org/vol17/iss1/art14/.

[11] Carpenter, S. R., Brock, W. A. and Ludwig, D., 'Collapse, learning, and renewal,' in Gunderson, Lance H., Craig Reece Allen, and Crawford S. Holling, eds. *Foundations of ecological resilience.* Island Press, 2012.

[12] Oba, G. *Climate change adaptation in Africa: A historical ecology.* Routledge, 2014.

[13] Roe, E., Untsinger, L H. and Labnow, K., 'High reliability pastoralism,' *Journal of Arid Environments* (1998) 39: 39–55.

Historical Research Methods

The methods used in this study might appear random with limited progressive linkages, partly because different historical periods and different types of data have been designated to describe adaptive strategies. To avoid such a perception, we will endeavor to thread together the variety of methods by thematic areas to highlight important aspects of adaptive disaster management. We use long-term historical methods to capture rare events that might be missed if only short-term studies were to be considered. It is also essential to identify the frequency of returning cycles of triggers that drove processes of collapse and resilience.[14] We present the evidence in the subsequent chapters, but meanwhile, it is worthwhile to explain how such disaggregated data collection methods may report resilience or collapse in pastoralism. One significant advantage of this study is that we use the same society in a particular geographical space, reflecting a marked change in their adaptive survival strategies. We consider internal changes and external influences that cause an organization to modify and transform its adaptive strategy.

Accordingly, data from long-term studies recognize evolving adaptive strategies helpful in assessing and evaluating pastoral resilience, collapse, reorganization, and transformation. Our 'threads' of the overall argument linking space and time comprise historical and contemporary processes of pastoralists' adaptive strategies at regional and longitudinal local scales. We emphasize how historical functions are merged into the modern, changing the trajectory of pastoralism as a way of life from the pre-colonial, colonial, and post-independence periods. The sequences disclose how planned adaptive activities are transformed through successive collapses of traditional pastoralism, land-use transformation, and an apparent weakening of older resource tenure systems into a semi-private agropastoralism. The shift, including livelihood activities, comprises new adaptive strategies.

Consequently, while the questions asked might have changed, the people concerned are the same. In terms of post-independence time, we merely scaled downwards the analysis from regional to local—in which case we are dealing with a local grouping of the Borana—which in the current study are referred to as the Obbu Borana—residents of the southern

[14] Strayer, D., Glinzenstein, J., S., Jone, C. G., Kolasa, J., Likens, G. E., McDonnell, M. J., Parlar, G. G. and Pickett, S. T. A., *Long term ecological studies: An illustrated account of their design, operation, and importance to ecology*, Occasional Publication of the Institute of Ecosystem Studies, (New York: New York Botanical Garden Millbrook, August 1986).

Ethiopia–northern Kenya borderland—administratively in the Sololo sub-county. The local processes we describe are connected with the region across the frontier. Our dialogue is about how regional environmental and political events influence the local operations of change. The synthesis chapter (Chap. 10) would make these ties more obvious.

Nonetheless, we need to explain the data types, albeit briefly. For the earlier sections, the work draws primarily on indigenous histories and social memory of adaptive responses to changing climatic conditions and cycles. Archival sources provide the foundation for dealing with the changes imposed by colonial overrule and national boundaries that dissected the Borana Oromo territory. Personal experience and interview data provide the core methods for measuring transformation for post-colonial statehood and administration's early and later phases. For the local cases, the same households were repeatedly interviewed and monitored for a decade or more on changes in individual responses to drought adaptations, while at institutional levels, families and individuals were repeatedly interviewed on differences in household wealth, social capital networks shared with neighborhoods in terms of the progression and transformation of social capital networks. Inevitably, these different data sources provide differing degrees of temporal and spatial specificity and resolution. Each source type is better suited to answering specific questions than others, as reflected in the planned content of the different chapters—that might be treated as sequential—cascading from a bigger picture at the regional to point-by-point details at community levels.

Accordingly, these different sources provide differing degrees of temporal and spatial specificity and resolution in answering specific questions. We may use an example of collapse here. Measuring failure in pastoralism requires a variety of tools and methods. In extreme cases such as those of the late nineteenth century, collapse in human demographics and livestock populations ruined pastoralism. In later periods, where state interventions have reduced collapses in human demographics through interventions (such as providing relief food), collapse in pastoralism infers crushes in livestock demography that adversely affects herd reproductive capacity. Such a depleted herd population would take a decade or more to recover when favorable conditions return. This system may, however, be hugely transformed. Such factors as fragmentations of grazing lands by farms and range enclosures would influence the extent to which the agropastoral economy is affected by recurrent droughts. This is at the economic level. At the social level, the direct consequence of the collapse in pastoralism results in shifts in household wealth—it means that under collapsed livestock demography, households will lack sufficient sources of livelihood. This has two consequences. First, society seeks alternatives by seeking non-pastoral livelihoods. Second, the collapse in household wealth, depending on the extent of impoverishment, results in the loss of the capacity to participate in social security networks for the members. This would be measured in terms of ratios of stock per capita, establishing a bare minimum under which the system is operative or fails, and livestock life histories of individual households. When household wealth collapses, the consequence is a loss of institutional capacity to conduct traditional methods of stock sharing that trigger the society to transform the system. Different resource-sharing rules from conventional pastoralism will be used in the transformed social capital networks. Accordingly, the multilayered effects of collapse would demand the development of other methods that refine assessment techniques in adaptive strategies. In all these changes, climate variability has remained a constant factor influencing the whole system at the scale of geographical space where actions take place. We will apply these ideas in the context of the geographical space.

Geographical Space

The region that is the focus of this study extends from the plains of Liban across the Dawa River in the north and central plateau of Dirre, extending to an escarpment (in southern Ethiopia) that marks the frontier with northern Kenya (see Fig. 2.2). The central Borana plateau has nine clusters of ancient well systems, whose engineering skills and technologies are associated with ancient hydraulic societies—the ancestors of the Borana Oromo.[15] The region is climatically and ecologically diverse.

The southern Ethiopian and northern Kenyan border is marked by the impressive physical features of the Gorro escarpment, extending for more than 600 km. The escarpments' sub-humid landscapes served historically as water sources for people and livestock on both sides of the frontier. Captain Butler initially marked the boundary[16]—this initial border—left all the wells and springs on the Ethiopian side, denying access to the

[15] Oba, G. *Climate change adaptation in Africa: a historical ecology*. Routledge, 2014.

[16] See also KNA/PC/NFD4/3/1, a frontier agreement between H.B.M. government and emperor Menelik, 6 December 1907.

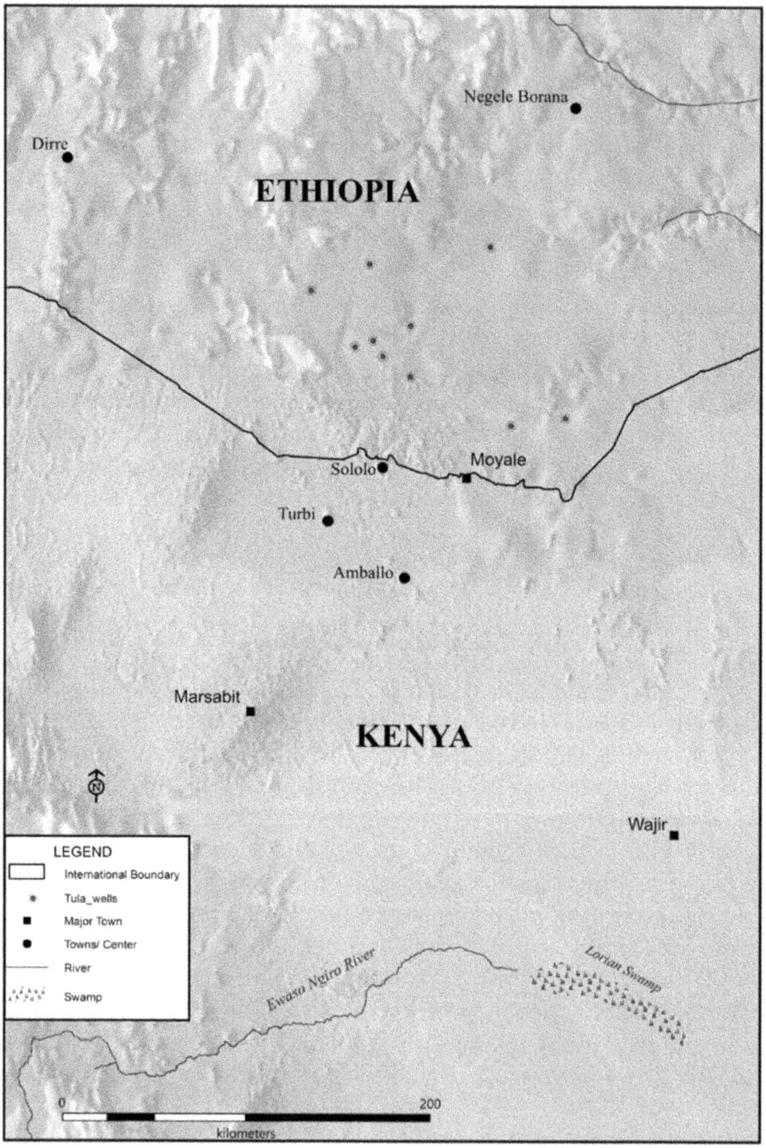

Fig. 2.2 Study region of the north and central plateau of Dirre extending to an escarpment (in southern Ethiopia) which marks the frontier with northern Kenya

pastoralists left on the British side. Despite Captain Gwynn's[17] later attempt to correct the imbalance distribution of resources by marking a second border, the Ethiopian Emperor Menelik objected to the adjustments.[18] Accordingly, the British were obliged to negotiate a pastoralist cross-border grazing treaty in 1907, designed to allow the British subjects to water and graze their livestock on the Ethiopian side.[19]

On the Kenyan side of the frontier, the escarpment forms a sharp ecological gradient marked by a semi-arid *Acacia siebreana* zone, creating an ecotone. Descending toward the Kenya side of the border, the agroecology changes to arid and very arid rangelands, which served as the core grazing lands. The two border communities share the institutions of the *Gada*-generation classes and membership in clan alliances and social capital networks, related through extended families and kith. The sub-humid zone escarpment was significant for another reason. It attracted Ethiopian highland farmers—the vanguard of Menelik's frontier army, who settled there and cultivated crops. It was these Ethiopian farmers who, in the 1960s and 1970s, influenced the transfer of farming technology to the Obbu Borana. Before 2010, the Obbu residents (currently estimated at between 20,000 to 30,000 people)—distributed across nine administrative locations (Fig. 2.3) crowded into a limited area (125 km^2)—along the narrow semi-arid ecotone zone developed two integrated systems of land use—crop cultivation and private range enclosures for conserving pasture.[20]

We describe the methods for evaluating adaptive strategies to past disasters outlined under the three thematic areas—connected by environmental and social events: regional memories of disasters (Chaps. 3 and 4), reorganization and adaptive diversities (Chaps. 5, 6 and 7), and collapse and transformation of social capital networks (Chaps. 8 and 9). Therefore, readers might appreciate the progressive changes over time and the shifts in the human capacity to deal with the modifications examined across the three thematic areas. We begin with using regional social memory as a historical research method.

[17] KNA/DC/MDA/4/2, Major Gwynn, 1909, Political records of Mandera District, NFD report, p. 22.

[18] Oba, G., *Nomads in the shadows of empires: contests, conflicts, and legacies on the southern Ethiopian-northern Kenyan frontier*. Brill, 2013.

[19] Oba, G. *Climate change adaptation in Africa*.

[20] Since the 1990s, the latter had expanded into the communal rangelands in the arid agro-ecological zone.

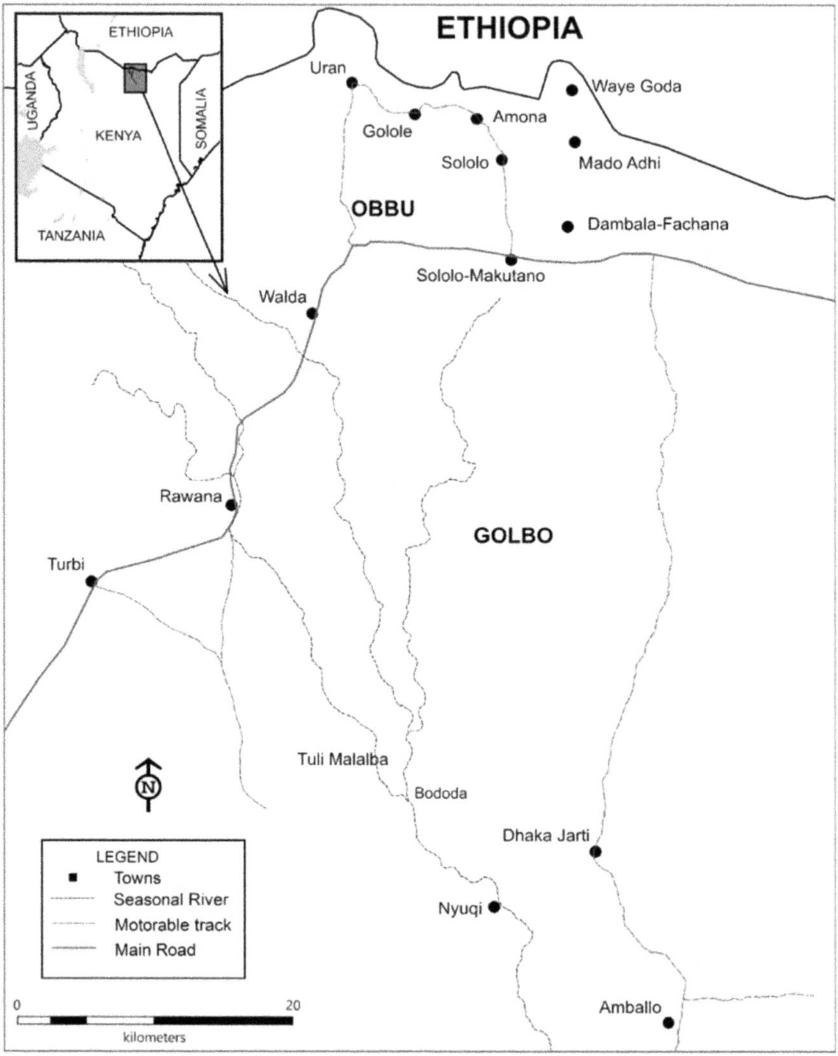

Fig. 2.3 Case study sites in the Obbu region of Northern Kenya

Regional Memories of Disasters

For those who experienced them, disasters seriously threatened human lives and economies (Chap. 3). The events are associated with social and political actions[21] that, when combined with proxy environmental indicators, are useful in analyzing regional-scale adaptations to disasters.[22] Knowledge of systemic failures in human adaptations helps to foretell the extent to which the integrity of a pastoral production system gets compromised. Further, understanding cataclysmic events embedded in social memory can contribute to constructing traditional event history calendars (EHCs).[23] Webster[24] has documented the value of social memory for reconstructing past disasters that induced population movements for the interlacustrine region of Central Africa. He provides matching evidence between social and environmental sources using known environmental proxy data. Additionally, social memory based on biographies of influential personalities, the transfer of power between generation classes, and cyclical names of generation classes helps reconstruct migrations, conflicts, and population collapses.[25] After disaster cycles or any crisis, building resilience relies on institutions that can reorganize society.[26] Thus, lessons can be learned by integrating oral traditions with written historical instruments. We now describe the use of social memory by the Borana Oromo.

The Event History Calendar

Among the Oromo, the inevitable cycle of events—the *daacci* cycle—is a tool for predicting the future. We interviewed oral historians to understand how they used the *gada* timeline, comprising cycles of generation classes (*gogessa*) and generation naming cycles (*maqbas*), to reconstruct an EHC. The time experts (*ayantu*) and oral historians described how

[21] Butzer, Karl W. 'Collapse, environment, and society.' *Proceedings of the National Academy of Sciences* 109, no. 10 (2012): 3632–3639.

[22] Hoffman, S. M. and Oliver-Smith, A., 'Catastrophe and culture: the anthropology of disaster,' *Anthropologica* 45 (1) (2003).

[23] Ali, S. M., *Huddur and the history of southern Somalia* (Cairo: Nahda Bookshop Publisher, 2005), 90–134.

[24] Webster, J. B., *Chronology, migration and drought in interlacustrine Africa* (London: Longman University Press, 1979).

[25] Oba, G., *Climate change adaptation in Africa.*

[26] Magis, K., 'Community resilience: An indicator of social sustainability,' *Social and Natural Resources* 23.5 (2010): 401–416.

disaster returned according to the *gogessa* and the *maqbas* cycles.[27] Using this approach, the Borana developed a traditional EHC for predicting the returns of megadroughts at intervals of 40,100—and 280-year cycles—when significant socio-political and economic changes are expected. The events were cross-checked with environmental proxy data to indicate regional climatic differences.[28]

From the EHC, we identified periods of megadroughts, famine, wars, locust plagues, and epidemics, with corresponding changes in human and livestock populations. We used the data from 1500 to 1900 to identify pastoral collapse and recovery periods. We discuss the inevitable return of history (*daacci* cycles) in predicting cyclical disasters over five centuries. We used the data to investigate the resilience of pastoralism during centuries marked by the vagaries of climate before colonial empires partitioned the region.

Using colonial archives and the region's climate history (1908–the 1960s), we analyze responses to political and climatic shocks (Chap. 4). We investigated the impacts of political frontiers between Ethiopia and Kenya on pastoralists' adaptations. We used extensive colonial archives on climate history to analyze how controlled grazing policies and variable climates influenced pastoralists' adaptive responses to disasters. This situation continued until the political independence of Kenya in 1963, when another collapse in pastoralism occurred, followed by reorganization and adaptive diversities.

Reorganization and Adaptive Diversities

Reorganization of adaptive diversities was concerned with trajectories of land-use change following the secessionist war (Chap. 5), individual herder drought management strategies (Chap. 6), and impacts of decadal droughts on livestock demographics and differences in household wealth (Chap. 7), which are summarized as follows.

[27] Legesse, A. *Gada: three approaches to the study of African Society* (New York: The Free Press, 1973); Tiki, W., Oba, G., Tredt, T., 'An indigenous time-related framework or reconstructing the impact of disasters on ancient water systems in southern Ethiopia, 1560–1950,' *Journal of Historical Geography* 41 (2013): 33–43.

[28] Tiki, Waktole, Gufu Oba, and Terje Tvedt. "An indigenous time-related framework for reconstructing the impact of disasters on ancient water systems in southern Ethiopia, 1560–1950." *Journal of Historical Geography* 41 (2013): 33–43.

Soon after Kenya's independence, the secessionist war collapsed pastoralism and transformed land use (Chap. 5). The secessionist war (*shifta*) displaced the pastoralists from their traditional grazing lands and forced them into security villages before dispersing again in the 1980s. Our research used the historical literature and extensive interviews of people (both men and women) across eight sites in the Obbu region (Fig. 2.3) to understand the long-term effects of the collapsed rural economy and processes of land-use change in terms of crop cultivation and rangeland enclosures. We describe the historical setting of farming and socio-economic adaptive strategies according to five pathways of land-use change and how each path transformed pastoralists' adaptive systems to disasters and changes in livelihoods.

We interviewed a total of 164 households randomly selected from eight locations. We interviewed individuals on their links with southern Ethiopian farmers in grain-clientele relations by outlining the history of grain reliance on Ethiopian settlers across the frontier. We examined the factors that compelled this society to participate in farming in a region that was, in fact, too marginal for crop cultivation. We further investigated and assessed factors that motivated participation in crop cultivation, the expansion of range enclosures, and the commercialization of grass production.

We determined methods of establishing farms and the number of years of farming. Since we were interested in the main constraints to agriculture, we identified and analyzed various indicators, including farming techniques, access to seeds, fertilizers, labor, and problems of weeds. Most importantly, we investigated people's perceptions of rainfall scarcity and irregularity and crops' resulting success or failure. We related rainy days and grain production and compared the effects of seasonal variations (short and long rains) on crop yields and long-term farm production by location. The role of droughts in farming cycles was investigated. At the household level, we determined grain self-sufficiency and food deficit months. We monitored crop production by case families for five years (1988–1992).

Additionally, we evaluated how the progressive expansion of range enclosures from the 1980s to the 2000s caused conflicts with non-enclosure owners over resource access rights. We investigated income-driven enclosed pastures and how this drives conflicts and future land privatization. Fragmentation of grazing landscapes—by crop cultivation and private range enclosures—adversely impacted individual herder drought management strategies.

We used individual biographies[29] of ten households to understand variations in herder drought management strategies (Chap. 6) between 1983 and 2011 on their livestock drought management strategies. The households—ranked into different cattle wealth classes—were monitored on their adaptive drought management strategies during seven drought years.[30] We asked the herders about their choices of direction and places of grazing movements, types of herd management, availability of labor, and herd losses.

We then investigated the impacts of decadal droughts on cattle demographics and changes in household wealth (Chap. 7). The initial survey 1983 focused on three sites (Sololo-Ramata, Mukutano, and Damballa Fachana). In 1991, five additional sites (Waye Godha, Mado Adhi, Anona, Golole, and Uran) were added to the sampling protocol. In total, we monitored 164 households over the period. During annual surveys, we requested families to recall cattle holdings in terms of calves (0–1 years), immatures (2–3-year-olds) (both males and females), reproductive females (4–12 years), and mature males (4–10 years) both before and after droughts. We converted the numbers into Tropical Livestock Units (TLUs), where adult females/males are equivalent to 0.7 TLU, immatures equal to 0.4 TLU, and calves weighted at 0.25 TLU. Total cattle heads were then multiplied by a weighted mean of 0.68 TLU. In relating cattle population dynamics and climate variability, we discuss details as follows.

The drought of 1983/84 was regional, causing massive stock losses.[31] In 1991/92, outbreaks of livestock epidemics resulted in high mortalities among cattle populations.[32] The droughts influenced structural changes in pastoral herd demography. Later, the severe flooding associated with the El Niño southern oscillation in 1997 and 1998 also impacted cattle populations following the outbreak of an epidemic of Rift Valley fever. This was

[29] Robert, F. 'The structure of autobiographical memory and the event history calendar: Potential improvements in the quality of retrospective reports in surveys.' *Memory* 6, no. 4 (1998): 383–406.

[30] The drought years were those of 1983/1984, 1991/1992, 1995, 1999/2000, 2006, 2008/2009, and 2010/2011.

[31] Xiaogang, S., 'Dynamics of continuity and change in pastoral subsistence among the Rendille in Northern Kenya: With special reference to livestock management and responses to Socio-economic change,' *African Study Monographs* 31 (2005): 1–94.

[32] Oba, G. 'The effect of multiple droughts on cattle in Obbu, Northern Kenya.' *Journal of Arid Environments* 49, no. 2 (2001): 375–386.

followed by yet another severe drought during 2001 and 2002 when poor short rains were followed by failure of the long rains.[33] Herd recovery was disrupted by the lack of 2006, widespread throughout the region.[34] The drought of 2011 was among the worst in the area.[35]

Herd structure in calves, immatures, breeding females, and males was monitored at different scales—by region, agroclimatic zones, and sites. Patterns of population change between the pre-drought of 1982 and 2011 were determined as population density differences between pre- and post-drought periods. Additionally, we evaluated the relationship between herd demographics and rainfall variability.[36] We used meteorological data from the Moyale station (80 km from Obbu) from 1980 to 2004. Across three agroclimatic zones (semi-arid, arid, very arid), rainfall is bimodal, with about 75 percent of the total precipitation falling during March–April–May (the long rains), with the remaining 25 percent occurring during October–November (the short rains). We related cattle holdings to long-term deviations of annual rainfall to explain shifts in cattle populations. Using regression analysis with the long and short rains, with a lag of one year, the departure from the long-term mean rainfall is related to cattle mortality. Total cattle mortality negatively impacted household wealth levels.

Dynamics of Household Wealth

Cattle wealth can be used to classify households into wealth ranks,[37] using cattle per capita and household herd sizes into the following wealth categories: destitute (<1 TLU), very poor (4.5 TLU), poor (4.5–9 TLU), medium (9–13.5 TLU), rich (13.5–18.0 TLU), very rich (18–22 TLU),

[33] Little, P. D., Mahmoud, H. and Coppock, D. L., 'When deserts flood: Risk management and processes among East African pastoralists,' *Climate Research* 19 (2001): 149–159.

[34] Berhanu, W., *Recurrent shocks, poverty traps and the degradation of the social capital base of pastoralism: A case study from southern Ethiopia*, Center of Evaluation for Global Action Agriculture for Development, Paper No. AFD-0916, issued in July 2009.

[35] McPeak, J. G. and Barrett, C. B., 'Differential risk exposure and stochastic poverty traps among East African pastoralists,' *American Journal of Agricultural Economics* 83 (3) (2001): 674–679.

[36] Ba, A., Lesnoff, M., Poccard-Chapuis, R., Moulin, C-H., 'Demographic dynamics and off-take of cattle herds in southern Mali,' *Tropical Animal Health Production* 43 (2011): 1101–1109.

[37] Thornton, P. K., Galvin, K. A. and Boone, R. B., 'An agro-pastoral household model for the rangelands of East Africa,' *Agricultural Systems* 76 (2003): 601–622.

and wealthy (> 22 TLU). Wealth ranks were determined after the droughts of 1983/84, 1991/92, 1995/97, 2000/1, 2005/6, and 2010/11. Changes in the levels of household wealth compared by climatic zones, sites, and years during the pre- and post-drought periods of 1982/83 and 1991/92 and the recovery year (1996) were determined. Additionally, we monitored the wealth life histories of six households between 1991 and 2003 from one of the sites (Anona) to determine the proportion of time that individual families shifted between wealth ranks. We then measured the recovery of pastoral wealth according to herd accumulation during runs of good years and herd collapses during periods of droughts.[38] The impact on household cattle holdings directly affected the collapse and transformation of clan social capital networks and the sustainability of women's neighborhood household food security networks.

Collapse and Transformation of Social Capital Networks

Three social capital network systems—one operated exclusively by men, the second run by both men and women (Chap. 8), and the third used solely by women (Chap. 9)—were investigated. The first type of social capital network is the traditional clan social capital network called *buusa gonoofa*. The sub-clans shared social capital networks in southern Ethiopia and northern Kenya.[39] In the second type of social capital networks, the performance of individual networks for sharing milk cows (*daabare*) practiced by relatives, kin, and bond friends was also evaluated. Due to the collapse of the pastoral economy, the indigenous clan social capital networks were later transformed into broad-based neighborhood social capital networks (*baalitti*) that took advantage of new economic opportunities (Chap. 8). The transformation of broad-based neighborhood networks incorporated the residents in Obbu.

[38] McPeak, J., 'Individual and collective relationality in pastoral production: Evidence from northern Kenya,' *Human Ecology* 33 (2005): 171–197.

[39] A parallel study forms PhD Thesis research by Boku Tache Dida, awarded by the Norwegian University of Life Sciences under research funding of the senior author between 2004 and 2008. boku_fulltext_thesis-with-cover-page-v2.pdf (d1wqtxts1xzle7.cloudfront.net).

Ethnography of the Borana Clan Social Capital Networks

The functioning of clan social capital networks would only be understood with information on the political and social organizations of the Borana.[40] Figure 2.4 is a simplified representation of the structure and administrative levels of the clan social capital networks. The hierarchical social obligations require coordination by political offices of the senior councilor (*hayyu*) elected to represent the interests of a clan in the *gada* council during their eight years in office (called *luba*).[41]

There are also ritual leaders (*Qaallu*) resident on the Ethiopian side of the frontier, representing the two moieties (the *Sabbo* and *Goona*). The coordination of clan-wide social capital networks remains with the *hayyu*, who appoints clan political office holders on both sides of the border. The

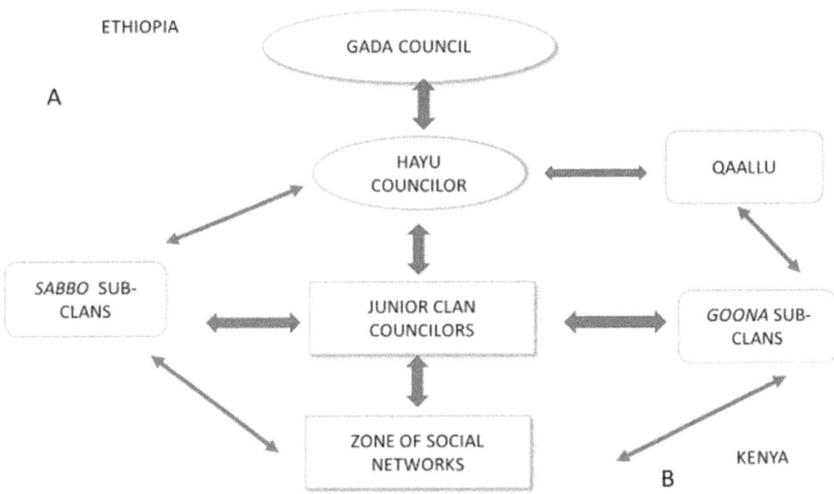

Fig. 2.4 The political structure of the Borana Oromo institutions and zones for supporting clan-based social capital networks (*buusa gonoofa*). The letter A represents communities in the "host" zones on the Ethiopian border, and B represents the "home" zones in Obbu

[40] For details of Borana social political organizations see Baxter, Paul Trevor William. *The social organization of the Galla of Northern Kenya*. PhD diss., University of Oxford, 1954.

[41] Legesse, Asmarom. *Gada: Three approaches to the study of African society* (The Free Press, 1973).

social capital networks on the Ethiopian side of the frontier were divided into geographically distributed zones (*gaadisa*—assemblies) that remained semi-independent shared networks with the Obbu region on the Kenyan side of the border that represents a single social capital network zone. The *hayu* councilors administer the clan social capital network at the regional level, assisted by the senior sub-clan leaders (*abba Qae*) and the junior councilors (*jallaba*)[42] who represent the sub-clans at localities. The common customary law of helping people experiencing poverty defined the criteria for who qualified for giving and receiving stock.

The Obbu region forms a single social capital network zone (*gaadisa*). At the same time, the Ethiopian side of the frontier has a series of overlapping social capital network zones linking people who belong to the same clan. The Obbu region represents the 'home,' while the social capital network zones on the Ethiopian border serve as 'hosts.' Each of these relationships forms a web of social capital networks bearing different obligations and responsibilities, expressed in terms of the rights of individuals. Individual networks for sharing milk cows (*haamesa*) also exist, negotiated by individuals without involving clans. The broad-based neighborhood networks have also emerged, part of a formal system created by the community. The participants include NGOs, legal groups, members of the community living in distant places in Kenya, and those in the diaspora. The changes created material shifts from the cattle economy to a financial one involving monetary contributions. We were interested in understanding the performances of clan social capital networks and why it was transformed into a broad-based system.

The process of stock sharing involves four steps. In the first step, between 1996 and 2005, we interviewed 25 elder councilors (*jallaba*) and senior sub-clan leaders (*abba Qae*) responsible for the welfare of the sub-clans. We used the narratives of the *jallaba* councilors to discuss why the system gradually lost its capacity to cope with dire socio-economic stress. We asked the junior councilors' opinions about the performance of the traditional *buusa gonoofa* social networks. In the second step, between 1991/92 and 2011, we randomly selected 104 households[43] from eight

[42] The *jaalab* who neglects the poor will be accused of abandoning their responsibilities, causing the destitute to starve. If found guilty, the *jaalab* is liable for paying a fine of five cattle to the sub-clan, and if he fails to be remorseful, he will be removed from office by having his authority revoked.

[43] This is part of the total sample.

locations and interviewed them about their participation in clan and individual social networks. We considered the relationship between herd sizes and the capacity to participate in social capital networks. We also determined (using correlation analysis) to what extent the clan and individual social capital networks had motivated individuals to participate. In the third step, we used records from 2002 to 2012 of nine neighborhoods in the peri-urban center of Sololo-Ramata to identify public fundraising initiatives for assisting needy households. We also investigated what specific changes led to the transformation of clan social capital networks into broad-based neighborhood networks. We used records of funds raised and the number of participants to evaluate the effectiveness of neighborhood household social networks. The neighborhood networks are gendered.

Gendered Social Capital Networks

We investigated how and why women's social capital networks maintained resilience (Chap. 9). We describe women's social security networks (called *maaro*) that also mitigate risks of environmental shocks and cover avenues inaccessible to men. Sharing among women involves the transfer of food items and other services, thus establishing relationships of giving and receiving. The obligation of *maaro* is between rich and poor households and among households of equal economic status. We used narratives of women to understand the structure and functions of women's social networks.

In Sololo-Ramata, we used individual social networks shared with the Catholic Church and the Mission Hospital. Between 1995, 1996, and 2005, surveys of five representative households sharing neighborhood networks with the Catholic Church and the Mission Hospital were requested to identify their sources of food security. Further, we compared women's social capital network operations in rural (Dambala Fachana) and peri-urban (Ramata) households. Between January 1995 and July 1996, we interviewed 99 families on their participation in these *maaro* networks. In the rural and peri-urban social networks, the operations of the networks were described in terms of the formation of formal women's social capital networks. We considered the functions of traditional institutions in terms of their active participation and items and services shared through public involvement. We then examined the roles played by dairy and grain markets in income generation and the extent to which rural and peri-urban women participate in credit-clientele systems at food shops and kiosks in supporting neighborhood household food security.

Additionally, to investigate weekly quantities and monetized values of shared food, in 1987, a sample of nine households in the Golole settlement were randomly selected, representing poor, medium, rich, and wealthy families. All nine households practiced livestock management and crop cultivation. During the weekly surveys, we asked women to recall what they had received the previous day from friends and the amounts they had shared. We assumed that the nine households were part of wider *maaro* social networks, with intra-communal food security supported by neighborhood household networks. The food items transferred from one household to another (but not consumed on the spot) tended to be minute. The amounts calibrated (using local miniaturized units such as spoons, bottle tops, and palm of hand) are then expressed in monetary values. The value of the food 'received' or 'shared' (compared with food directly purchased) provided a measure of the household's level of food security. The research in Chap. 10 synthesizes vital emerging issues from the historical study and makes concrete proposals on future mitigations of droughts based on the findings of Chaps. 3, 4, 5, 6, 7, 8 and 9.

CHAPTER 3

Social Memory on a Historical Scale: Configuring Cyclic Disasters, 1500–the 1900s

How might traditional, contemporary populations relate to past disasters without written records? This question is relevant for diverse African societies. Yet, there is evidence that socio-political institutions that set a particular organization apart might show clear continuity between past and present. Such institutions can serve posterity by passing life experiences orally from one generation to another through 'social memory' that describes past events in detail.[1] The past can form a basis for creating a historical calendar familiar to the present generation.[2] This chapter combines social memory and proxy environmental records to reconstruct pastoralist societies' resilience, collapse, and reorganization over five centuries, focusing on the Borana Oromo as a case study. The material contextualizes the history of pastoralism in the region, providing a regional framework to reconstruct the social memory that pre-dates colonial contacts. In those times, societies existed as herders and

[1] Tiki, W., Oba, G. and Tvedt, T., 'An indigenous time-related framework for reconstructing the impact of disasters on ancient water systems in southern Ethiopia, 1960–1950,' *Journal of Historical Geography* 41 (2013): 33–43; Oba, G. *Climate change adaptation in Africa: An historical ecology* (London: Routledge, 2014); *Herder warfare.in East Africa: Social and spatial history* (Winwick, White Horse Press, 2017).

[2] Miller, D. R., Steele, C. R., Bilgen, I. and Belli, R. F., 'Health status measurement via event history calendars vs. Conventional questionnaires: long time, no status' *AAPOR* May 14–17, 2009.

G. Oba, *Pastoralist Resilience to Environmental Collapse in East Africa since 1500*, https://doi.org/10.1007/978-3-031-48291-5_3

foragers, shifting between livelihood types depending on the extent of social and economic collapses associated with historical disasters. We examined specific disasters the case community faced and how they responded to them.

In this work, we use a variety of disasters as historical event markers, including severe droughts that caused famine, pandemics that decimated human and livestock demography, and wars that caused population displacements. We examined specific disasters and events included in the Borana event history calendar. We investigated how socio-political institutions serve as a repository for social memory and how crises associated with certain disasters transform human adaptations over five centuries.

This chapter has three goals. First, it shows how adaptive strategies to historical disaster events have transformed the practice of pastoralism in the present-day region of southern Ethiopia and northern Kenya. Second, it uses social memory and environmental proxies to evaluate pastoralists' adaptive strategies on a regional scale. Third, using the social memory of the Borana Oromo, it sets out to trace past historical disasters and adaptations enacted by that society. Our timeline is based on the Oromo *gada* (generation classes)[3] for reconstructing environmental and social history using an event history calendar (EHC). Results of a traditional event history calendar are used to interpret cycles of disasters from the sixteenth to the twentieth centuries following the *gada* generation class and the seven naming *maqbas* cycles (see below).

The chapter is divided into five sections. The first section summarizes potential applications of published proxy environmental indicators with social memory for reconstructing the impacts of past climates on pastoral populations. The second section configures a framework for analyzing the historical material and laying the foundation for integrating historical methods with oral traditions. The third section presents the Borana Oromo social memory on a historical scale. The fourth section presents the results of a traditional event history calendar and environmental proxy indicators. The fifth section assesses the restoration of pastoral resilience.

[3] Legesse, A. *Gada: Three approaches to the study of African society* (New York: Free Press, 1973).

Application of Environmental Proxy Indicators and Climate History

African pastoralism is the product of climate change.[4] In a highly flexible land-use system that shifted between swings in climate—from wetter episodes to drier periods, environmental historians and archeologists have established that this ancient land use has developed social memories of climate footprints in their oral narratives. Accordingly, the best method of understanding past climates and their impacts on societies is to analyze historical social vulnerabilities to droughts and famines to know how their knowledge systems of cyclic climate changes were registered in social memory.[5] It allows historians to decode the broader trajectories of changes, their consequent influence on the populations, and how changes in climate events at various scales left their imprints in social memory.[6]

For historians working in East Africa, where climate records earlier than the nineteenth century are limited, oral history and environmental proxy data have been standard tools for reconstructing the impacts of past climate on ecosystems.[7] For example, East African lakes have been the sources of rich paleoenvironmental 'archives' used as proxy data for reconstructing climate variations over several centuries.[8] The paleo-proxy environmental indicators show high sensitivity to terrestrial ecosystem changes and land uses.[9] Accordingly, it has been possible to reconstruct centuries of climate variations with droughts and famine interspersed with wet periods using sediment cores and fossil pollen. These periods of hunger were etched into the social memories of pastoral populations across East Africa.[10] As reported in the regional literature, the disasters were responsible for

[4] Oba, G., *Climate change adaptation in Africa: An historical ecology* (London, Routledge, 2014).

[5] Lamb, H. H., *Climate history and the modern world* (London, Routledge, 1982).

[6] Olson, J. M., Alagarswamy, G., Andresen, J. A., Campbell, D. J., et al., 'Integrating diverse methods to understand climate-land interactions in East Africa,' *Geoforum* 39 (2008): 898–911.

[7] Hartwig, G., W., 'Oral data and its historical function in East Africa,' *The International Journal of African historical Studies* 7 (3) (1974): 468–479.

[8] Verschusen, D., 'Lake-based climate reconstruction in Africa,' *Hydrobiologia* 500 (2003): 315–330.

[9] Ssemmanda, I., Gelorini, V. and Verschurren, D., 'Sensitivity of East African savannah vegetation to historical moisture-balance variation,' *Climate Past* 10 (2014): 2067–2080.

[10] Nicholson, S. E., 'Climatic and environmental change in Africa during the last two centuries,' *Climate Research* 17 (2001): 123–144.

extensive human fatalities associated with the collapses of rural economy and human demography.[11] For the remaining parts of the chapter, we focus on how environmental historians combined social memory and proxy ecological data to evaluate the long-term impacts of climate variability on the environment and societies in the context of environmental and social change.

Configuring Framework for Social Memory

Jan Vansina[12] likens social memory to a library; that is, social memory experts 'label' historical events so that they can be traced by recalling the event, like borrowing a book from a library. Although generally restricted to the society concerned, social memory can be disseminated to the public—in the past and the present.[13] Social memory is based on reliable sources for oral transmissions, such as older experts, despite gaps that may exist in their memories.[14] This being the case, the timing of oral histories is associated with the generation class in power at any one time, serving as a social and political change marker. Considering these institutions are dynamic, the events and their returns through generations are cyclic, developing a sense of the past returning in the future.[15] These cyclic events reflect schemes for memorizing past disasters and predicting future ones. Not only do they join the past to the present, providing a yardstick of historical events, but they also define several types of adaptive strategies.[16] The events are usually associated with geographical spaces, building

[11] Slavin, P., 'Climate and famines: a historical reassessment,' *WIREs Climate Change* 7 (2016): 433–447.

[12] Vansina J., 'Memory and oral tradition,' in Miller, J. C. (Eds.) *The African past speaks: Essays on oral tradition and history* (Hamden: Archon, 1980).

[13] O'Leary, M. F., 'Drought and change amongst northern Kenya nomadic pastoralists: The case of the Rendille and Gabra,' in Palsson, G. (Ed.) *From water to world-making* (Uppsala: The Scandinavian Institute of African Studies, 1990).

[14] Tiki, Waktole, Gufu Oba, and Terje Tvedt. "An indigenous time-related framework for reconstructing the impact of disasters on ancient water systems in southern Ethiopia, 1560–1950." *Journal of Historical Geography* 41 (2013): 33–43.

[15] Cukierman, D. and Delgrande, J. P., 'Expressing time intervals and repetition within a formalization of calendars,' *Computational Intelligence* 14.4 (1998): 563–596.

[16] Breshears, David D., López-Hoffman, L. and Graumlich, L. J., 'When ecosystem services crash: preparing for big, fast, patchy climate change.' *Ambio* 40.3 (2011): 256–263.

memory themes among societies.[17] The mechanisms for building the infrastructure for social memory include oral poetry, folklore, livestock grazing movements, geographical areas, and recalling personalities of importance. Regarding geographical scales and places that define the events, the physical spaces provide effective sources of memory for individual autobiographical experiences—such as names of persons and references to age sets during historical periods.[18]

The challenge is retrieving oral knowledge using memories embedded in social institutions and interpreted by historians. Apart from informative linguistic history about earlier periods,[19] there are few historical studies of nomadic peoples in East Africa before the nineteenth century.[20] David Henige[21] suggests specific actions as to how historians might use oral knowledge where written sources are lacking or as a way of cross-referencing the reliability of a particular method. We advocate for developing a better understanding of traditional systems used by 'non-calendrical' societies. Such an approach would demand a sense of where, why, and how conventional explanations of events could be of value to historians.

The first necessary action is to build sequences in the system by organizing critical events into a chronological calendar (this might be approximated by the coming to power of generation classes and age sets).[22] Since local institutions tend to transmit knowledge of important traditional events while ignoring less important ones, the chronological list of events might miss incidents of historical significance. Nonetheless, as Henige[23]

[17] Oba, G. *Herder warfare in East Africa; Spatial and social history* (White Horse: xx, 2017).

[18] Belli, Robert F. 'The structure of autobiographical memory and the event history calendar: Potential improvements in the quality of retrospective reports in surveys.' *Memory* 6.4 (1998): 383–406.

[19] Ehret, Christopher. 'Historical Linguistics.' *The International Encyclopedia of Anthropology* (2018): 1–9.

[20] Van Zwanenberg, R. M. A. and King, A., *An economic history of Kenya and Uganda 1800–1970* (London: Macmillan, 1975); Waller, R., 'Ecology, migration, and expansion in East Africa' *African Affairs* 84 (336) (1985): 347–370; David M. Anderson, eds. *Resilience and Collapse in African Savannahs: Causes and consequences of environmental change in east Africa* (London: Routledge, 2018); Samuel F. Derbyshire, *Remembering Turkana: Material Histories and Contemporary Livelihoods in North-Western Kenya* (London, Routledge, 2021).

[21] Henige, D. P., *The chronology of oral tradition: Quest for a chimera* (Oxford: Clarendon Press, 1974).

[22] Bollig, Michael. 'Adaptive cycles in the savannah: pastoral specialization and diversification in northern Kenya.' *Journal of Eastern African Studies* 10.1 (2016): 21–44.

[23] Henige, D. P., *The chronology of oral tradition*.

contends, '[t]he chronological content of oral tradition …may be useful in assessing the duration of the past and the sequence of events within it.' Consequently, '[l]ooking at traditions through the eyes of African oral narrators…yields a topology of narratives based on the distinction between absent past and present past that also demarcates the sorts of historical referent predominant in each category.'[24] This construct is momentously important for pastoral societies, whose mobile land-use practices help them connect landscapes and history. This applies not only to Neolithic pastoralists but also to contemporary ones who leave behind evidence of material valuable culture for dating their presence.[25] Among modern pastoralists, each landscape offers proof of past human memory, earlier settlements, and names—some of which describe the geographical uniqueness of historical events. In diverse types of analysis, time frames might be vague or represented symbolically in terms of cultural histories and social memory prompted by different landscapes, thereby linking landscapes with traditions, history, and social memory.

We frame social memory regarding environmental history (EH), which examines how environmental changes influence human actions and social behavior. David Anderson[26] argues that connections between past and present memories link the environment and societies by influencing human behavior in response to environmental events. For Emmanuel Le Roy Ladurie[27]—whose thoughts on the functions of 'deep time' in directing historical arguments have significantly influenced the scholarship—different conclusions might be reached depending on the analysis discipline. Indeed, the 'deep time' perspective enables insights into the causes of systems collapse in the long term and the choices available to affected populations in the short term framed in terms of historical framework.[28]

[24] Miller, J. C. 'Introduction: listening for the African past,' in J.C. Miller (Ed.), *The African Past Speaks: Essays on Oral Tradition and History*, Folkestone, 1980.

[25] Lane, P., 'Trajectories of pastoralism in northern and central Kenya: An overview of the archeological and environmental evidence,' in Bollig, M., Schnegg, M and Wotzka, Ha-P. (Eds.) *Pastoralism in Africa: Past, present and future* (New York: Berghahn, 2013).

[26] Anderson, D. M., 'The beginning of time? Evidence for catastrophic drought in Baringo in the early nineteenth century,' in Anderson, D. M. and Bollig, M. (Eds.) *Resilience and collapse in African savannahs: Causes and consequences of environmental change in East Africa* (London: Routledge, 2017).

[27] Le Roy Ladurie, E., *Times of feast, times of famine: A history of climate since the year 1000* (New York: Garden City, 1971).

[28] Redman, Charles L. and Kinzig, A. P., 'Resilience of past landscapes: resilience theory, society, and the longue durée.' *Conservation ecology* 7.1 (2003).

Accordingly, we use history as an 'available guide.' This means that we present direct and indirect evidence of how past environmental changes influenced the adaptive strategies of societies. Caviedes[29] evaluates the effects of past events as explained by contemporary models by examining how disasters such as droughts, excessive rainfall, and pandemics precipitated collapses in rural populations and economies. Using direct and indirect evidence, historians try to make sense of the connectedness of disasters and resulting social and economic survival strategies.[30] This point of view does not suggest that all observed changes are associated with natural events—instead, others might be triggered by external and internal social and political forces.[31]

Critical events might be local, regional, or both. At both scales, the complex relations between the environment and societies created 'traditions' that specialize in past experiences recalled by the present generations.[32] Since the historical past serves as a 'library' for understanding climatic adversities, it is possible to gauge, for example, how the changes influenced adaptive strategies. From various analyses, it is possible to identify individual and community behaviors affecting social responses.[33] Interestingly, some narratives of African pastoralist societies have identified groups that had 'disappeared,' yet they continue to exist in the form of historical memories.[34] During periods of crisis, the unraveling of social structures that generally bind a society together meant that the capacity to help groups survive was lost. Eventually, the breakdown in internal

[29] Caviedes, C. N., *El Niño in history. Storming through ages* (Florida, FL: University Press of Florida, 2001).

[30] Namias, J., 'Severe drought and recent history,' in Rotberg, Robert I., and Theodore K. Rabb, eds. *Climate and history: studies in interdisciplinary history*. Vol. 341. (Princeton University Press, 2014).

[31] Rabb, T. K., 'Climate and Society in History: A Research Agenda,' *Social Science Research and Climate Change* (1983): 61–76.

[32] Naylor, S., 'Historical geography: Natures, landscapes, environments,' *Progress in Human Geography* 30.6 (2006): 792–802.

[33] Adger, W. Neil, Suraje Dessai, Marisa Goulden, Mike Hulme, Irene Lorenzoni, Donald R. Nelson, Lars Otto Naess, Johanna Wolf, and Anita Wreford. 'Are there social limits to adaptation to climate change?' *Climatic change* 93 (3) (2009): 335–354.

[34] Sutton, J. E. G. 'Becoming Masai land' in Spear, T. and Waller, R. (Eds.), *Being Maasai: Ethnicity and identity in East Africa* (London: James Currey, 1993); Martini, I. P. and Chesworth, W., *Landscapes and societies* (Dordrecht: Springer, 2010).

adaptive strategies disaggregated societies, causing them to lose territories they had relied on for survival.[35]

The collapse of livestock and human demography resulted in the abandonment of tribal grazing territories. In the process, they lost not only wealth but also their identities. Some societies that collapsed and finally became extinct as an entity may have prevailed until that final stroke—a combination of adverse climatic and economic conditions, pestilence, and conquests by more powerful groups expunged them except for historical memories.[36]

We present an example here. James Webb's[37] exhaustive analysis of the Sahelian region from 1600 to 1845 describes how pastoral herders from distinct cultures responded to the shifting climate and social and political conflicts. They moved from the northern Sahara into the Sahelian and Soudian zones, and these shifts introduced different production systems. In his narrative, Webb uses the environment as an overarching driver of societal change. Economic and political motives drove the population shifts and changes in land use along the 'desert frontier.' Similarly, the Horn of Africa notably shares many social, political, and economic dynamics. Progressive population shifts also introduced conflicts with other groups—these further agents of disaster contributed to the collapse of pastoralism. The changes were not necessarily orderly since 'collapse' was due to demographic declines with multiple consequences. Instead, they were sporadic and chaotic, including complexities and uncertainties in regime shifts.[38] Historians try to understand available sources about how disasters changed social, ecological, economic, and political structures.[39] Social memory thus serves as a tool for reconstructing such past events, as we now discuss.

[35] Cremaschi, M. and Zerboni, A., 'Human communities in a drying landscape,' in Martini, I. P. and Chesworth, W., *Landscapes and societies* (Dordrecht: Springer, 2010).

[36] Martini, I. P., and Chesworth, W. *Landscapes and societies: selected cases.* Dordrecht: Springer, 2010; Waller, Richard. 'Ecology, migration, and expansion in East Africa,' *African Affairs* 84.336 (1985): 347–370.

[37] Webb, J. L. A., *Desert frontier: Ecological and economic change along the western Sahel 1600–1850* (Madison, WI: University of Wisconsin Press, 1995).

[38] Kinzig, A. P., Ryan P., Entienne,M., Allison, H., Elmqvist, T., and Walker, B. H., 'Resilience and regime shifts: Assessing cascading effects,' Ecology and society 11.1 (2006): 20 [online] URL: http://www.ecologyandsociety.org/vol11/iss1/start/20/.

[39] Holling C.S., Gunderson, L. H. and Ludwig, D., 'Inquest of a theory of adaptive change,' in Gunderson, Lance H., Craig Reece Allen, and Crawford S. Holling, eds. *Foundations of ecological resilience.* Island Press, 2012.

For the pastoral communities in northeast Africa, various disasters left marks on their adaptive strategies. The histories of disasters thus reflect the histories of changing adaptive systems by affected populations. Crises tend to be cyclic events that have transformed the survival strategies of pastoral peoples over centuries.[40] Because droughts are recurrent, pastoralist societies became preoccupied with making predictions using various indigenous methods (including astronomy, studying intestines of slaughtered stock, animal behavior, and the naming cycles of generation classes). Although these methods might be 'unscientific,' they helped pastoral societies to develop adaptive strategies for environmental disasters.[41]

Environmental disasters have physical, social, and economic dimensions that might occur abruptly or gradually. Disasters affect food sources, diminish resources on which societies depend, and trigger conflicts over limited resources. For pastoral communities in the past, every catastrophe that resulted in stock losses was followed by periods of ethnic strife as the affected populations tried to recoup their losses. One of the reasons is that famine induced by sustained disasters decomposed populations by drastically modifying their adaptive survival strategies.[42] Exploring past disasters shows processes of change experienced by societies and how the past would come to influence their future adaptations.[43] Similar to natural disasters, armed conflicts caused mortalities and property loss, while pandemics caused the collapse of human and animal populations.[44] Pandemics such as those in the nineteenth century reduced livestock populations,

[40] Ellis, J., 'Climate variability and complex ecosystem dynamics: Implications for pastoral development,' in Scoones, I. (Ed.) *Living with uncertainty: new directions in pastoral development in Africa* (1994).

[41] Hulme, M., Doherty, R., Ngara, T., New, M. and Lister, D., 'African climate change: 1900–2100,' *Climate Research* 17: 145–168.

[42] Cremaschi, M. and Zerboni, A., 'Human communities in a drying landscape: Holocene climate change and cultural response in the central Sahara,' in Martini, I. Peter, and Ward Chesworth, (Eds.) *Landscapes and societies: selected cases*, Springer Science & Business Media, 2010.

[43] Schenk, G. J., '"Learning from history"? Chances, problems and limits of learning from historical natural disasters,' in Krűger, F., Bankoff, G., Cannon Terry, Orlowshi and Schipper, L. E. (Eds.) *Cultures and disasters-understanding cultural framing in disaster risk reduction* (London: Routledge, 2015).

[44] Leaning, F. and Guha-Spir, D., 'Natural disasters, armed conflicts and public health,' *The New England Journal of Medicine* 369 (2013): 1836–1842.

with recoveries taking decades.[45] A remarkable constant in these discussions is the interaction of exogenous and endogenous factors that impact societies. Exogenous factors tend to alter the trajectory of change, while endogenous factors are the means organizations use to respond to external pressure points (see Chap. 4).

We have discussed disasters and their impacts from a broad historical perspective that will be useful for identifying pressure sources and points—each of which can disrupt traditional pastoral production. Theoretically, therefore, we have shown several of the consequences of external pressure points, such as the collapse of populations. Depending on the intensity, frequency, and duration of exogenous stressors, collapsed pastoral societies might use internal self-adjustments (endogenous factors) to reorganize or transform social institutions and livelihoods. These adjustments thus become the building blocks of social memory.

While disasters themselves depict memorable events, post-disaster periods are remembered because it was then that a society rebuilt economies and strengthened weakened social institutions.[46] In this way, the performance of pastoral economies may be assessed during intermediate periods during which social recovery occurred. These are the constructs of the traditional calendars. We will consider using traditional calendars that provide historians with innovative tools to establish the historical chronology of past events.[47] This chronology may be inferred from the perspectives of societies that had evolved social organizations, institutions, and territorialities to control responses to environmental changes.

Using traditional calendars, the pastoralists of northeast Africa have developed predictive methods for cyclical events. For example, the Somalis organized time in seven-year cycles, corresponding with weekly cycles—such as the Wednesday year (*arabaaco*), the Saturday year (*sabti*), and so on.[48] Among the Gabra and the Rendille, time is counted in sevens and

[45] Weiss, H., '" Dying cattle": Some remarks on the impacts of cattle epizootics in the central Sudan during the nineteenth century' *African Economic History* 26 (1998): 173–199.

[46] Tiki, W., Oba, G. and Tvedt, T., 'An indigenous time-related framework for reconstructing the impact of disasters on ancient water systems in southern Ethiopia, 1960–1950,' *Journal of Historical Geography* 41 (2013): 33–43.

[47] Anderson, D. M., 'The beginning of time? Evidence for catastrophic drought in Baringo in the early nineteenth century.'

[48] Massey, G., 'Subsistence and change,' in Massey, G. (Ed.), *Lessons of Agropastoralism in Somalia*. Boulder: Westview (1987).

can be used to reconstruct the event history calendar. Paul Robinson[49] has used the oral history of the Gabra to reconstruct event history calendars for a century. The Gabra recognize the return of regional drought and famine in cycles of seven years (*ogolti*) and thirty years (*daacci*). They also recognize the seventy-year cycles (*kontomata*) and eighty-to-hundred-year cycles (*adaal*).[50] According to these pastoral societies, the return of cyclical events is inevitable, which enables them to be vigilant in the future. We will next apply this framework to the social memory of the Borana Oromo.

The Borana Oromo Social Memory on a Historical Scale

The Borana Oromo, whose ancient territory was part of present-day southern Ethiopia and northern Kenya, their daily lives were strictly controlled by time and rituals. Asmarom Legesse[51] refers to them as people with 'the most ingenious creations of the human mind.' One of these creations he refers to is the Borana knowledge of time—using the *gada* timeline (generation class cycles). Using *gada* as a repository of social memory, the Borana have developed a historical calendar for the past 600 years regarding interrelationships between active *gada* cycles and significant historical, social, political, and environmental events. This makes the Borana among the few African societies with a unique and detailed knowledge of past times and circumstances.[52]

The Borana concept of cyclical times has been explained elsewhere[53] and is only summarized here. Each *gada* stays in power for eight years before handing control over to the next patri-classes (*gogessa*) into which

[49] Robinson, P.W., *Disaster and response among the Gabbra of northern Kenya: An historical perspective*, discussion paper for the conference on ecological stress in East Africa, sponsored by Northwestern University and the National Museums of Kenya, Nairobi, 15–17 June 1980, Unpublished manuscript.

[50] Oba, G., 'The importance of pastoralists' indigenous coping strategies for planning drought management in the arid zone of Kenya,' *Nomadic Peoples* 5 (1) (2001): 89–119.

[51] Legesse, A. *Gada: Three approaches to the study of African society* (New York: Free Press, 1973, 158, 179).

[52] For other East African pastoralists see, Odongo, O. K. and Webster, J. B., *The central LWO during the Aconya*, East African Literature Bureau, Nairobi: 1976.

[53] Tiki, W., Oba, G., Tvedt, T., 'An indigenous time-related framework for reconstructing the impact of disasters on ancient water systems in southern Ethiopia, 1560–1950,' *Journal of Historical Geography* 41 (2013): 33–43; Oba, G., *Climate change adaptation in Africa: An historical ecology* (London: Routledge, 2017), 125–136.

the society is organized. It takes another 40 years (5×8 years) before the same *gada* class returns to power for another eight years. The *gogessa* are clustered into seven naming cycles (*maqbas*),[54] which return in cycles of 7×40 years (280 years).

The Borana perception of time is that political, environmental, and social events repeat themselves during various *gada* cycles. This inevitable return of history is called *daacci*, which marks the return of the *gogessa* and *maqbas* cycles. The *gada*, *gogessa*, and *maqbas* cycles were used to reconstruct the traditional event history calendar (Table 3.1). In the unfolding *gada* cycles, the Borana provide methods for managing disasters and coordinating societal actions. Under extreme circumstances, the social-institutional continuum breaks down due to demographic and economic collapses. The traditional event history calendar captures these events.

In this case study, we used narratives of oral historians to reconstruct event histories corresponding with the returns of the *gogessa* and *maqbas* cycles of the Borana, beginning with disaster events during the sixteenth through the seventeenth, eighteenth, nineteenth, and twentieth centuries. The oral calendar was then checked against regional environmental proxy indicators (Table 3.1). Based on the calendar, we then discuss diverse types of disasters. We now examine cycles of megadroughts and famines, excessively wet years, locust plagues, external and internal conflicts, pandemics, and population collapses.

Megadrought and Famine Cycles

Oral historians mention periods of megadroughts, individuals that lasted for eight years. The recalled disasters are supported by written sources from the sixteenth through the nineteenth centuries.[55] Droughts predictably induced famine, resulting in high human and livestock mortality. Consequently, large areas of the grazing lands were left empty of human habitation.[56]

[54] The seven maqbas are: *Libasa, Sabbaqa, Makula, Fullassa, Mardida, Moggisa* and *Daraar*.

[55] Tiki, W., Oba, G., Tvedt, T., 'An indigenous time-related framework for reconstructing the impact of disasters on ancient water systems in southern Ethiopia, 1560–1950.'

[56] Lamphear, J., 'Aspects of becoming Turkana: Interactions and assimilation between Maa and Ateker speakers,' In Spear, T. and Waller, R. (Eds.) *Being Maasai: ethnicity and identity in East Africa* (Ohio University Press, 1993).

Table 3.1 Reconstruction of disasters using the gada (gogessa and maqbas) cycles for constructing the EHC of the Borana Oromo

Gada timeline	*Calendar*	*Maqbas*	*Disaster events*	*Sources of proxy data*
Dagale Yaya	1536–1544	*Libasa*	Famine	NMSA[a]
Asosa Titile	1544–1552	*Sabbaaqa*	Regional drought	Degefu Pankhurst[b]
Boru Luku	1552–1560	*Moggisa*	Megadrought	Pankhurst[c]
Abay Horoo	1560–1568	*Makula*	Normal	Schove[d]
Bidhu Dhoqqe	1568–1576	*Fullassa*	Excessive rainfall	Oral
Oroo Dullacha	1576–1584	*Mardida*	Normal	Oral
Yaya Horoo	1584–1592	*Darara*	Regional drought	Schove[e]
Doyo Boro	1592–1600	*Libasa*	Population migrations	Oral
Baco Nadha	1600–1608	*Sabbaqa*	Normal	Oral
Urgumesa Iggo	1608–1616	*Mogissa*	Droughts and famine	NMSA[f]
Babbo Horoo	1616–1624	*Makula*	Famine and epidemic	NMSA[g]
Babbo Sibu	1624–1632	*Fullasa*	Plagues of locusts	Oral
Indaale Doyo	1632–1640	*Mardida*	Plagues of locusts	Pankhurst[h]
Accu Abiyu	1640–1648	*Daraar*	Reduced population	Oral
Abbu Lakhu	1648–1656	*Libbasa*	Famine	NMSA[i]
Abayi Babbo	1656–1664	*Sabbaqa*	Population movements	Oral
Alle Kura	1664–1672	*Mogissa*	Ethnic conflicts	NMSA[j]
Wayu Reele	1672–1680	*Makula*	Warfare	NMSA[k]
Morowa Abay	1680–1688	*Fullasa*	Recovery, prosperity	Oral
Gobba Alla	1688–1697	*Mardida*	Famine	Oral
Dawwe Gobo	1697–1705	*Daraar*	Drought everywhere	NMSA[l]
Jarso Iddo	1705–1713	*Libbasa*	Wars and conflicts	Oral
Walle Waccu	1712–1720	*Sabbaqa*	Social disintegration	Webster Schove[m]
Sora Dadacha	1720–1728	*Moggisa*	Recovery, prosperity	Schove[n]
Dadacha Robale	1728–1736	*Makula*	Normal	Oral
Halakhe Doyo Arele	1736–1744	*Fullasa*	Social discontent[o]	Oral
Guyo Gedo	1744–1752	*Mardida*	Locust plague	Pankhurst[p]

(*continued*)

Table 3.1 (continued)

Gada timeline	*Calendar*	*Maqbas*	*Disaster events*	*Sources of proxy data*
Mad'a Boru	1752–1760	*Daraar*	Epic famine	Pankhurst[q]
Dadacha O. Morowa	1760–1768	*Libbasa*	Wars	Herring[r]
Bule Dadacha	1768–1776	*Sabbaqa*	Severe drought, wars	Pankhurst[s]
Liban Wata	1776–1784	*Mogissa*	Ethnic conflicts	Schove Herring[t]
Wayu Raale	1784–1792	*Makula*	Famine	Pankhurst[u]
Boru Mad'a	1792–1800	*Fullasa*	Strengthening of *gada*[v]	Wood[w]
Ungule Lakhe	1800–1808	*Mardida*	Excessive rainfall	Degefu[x]
Saaqo Dadacha	1808–1816	*Daraar*	Famine and drought	Oral
Jilo Nencoo	1816–1924	*Libasa*	Normal	Schove[y]
Sokhore Anna	1824–1832	*Sabbaqa*	Famine	Degefu NMSA[z]
Madha Boru	1832–1840	*Moggisa*	Famine, epidemic	Wood Håkansson[aa]
Liban Jilo	1840–1848	*Makula*	Internal conflict[bb]	Oral
Jaldesa G. Dabasa	1848–1856	*Fullasa*	Ethnic conflicts, famine	Dale[cc]
Doyo Jilo	1856–1864	*Mardida*	Internal conflict	Oral
Haro Adi	1864–1872	*Daraar*	Internal conflict	NMSA[dd]
Dida Bitata	1872–1880	*Libasa*	Internal conflict, famine	Hastenrath[ee] Ssemmanda[ff]
Guyo Boru	1880–1888	*Sabbaqa*	Internal conflict	Oral
Liban Jaldesa	1888–1896	*Moggisa*	Rinderpest epizootic, wet	Wood[gg] Conway[hh]

The oral sources are checked against environmental proxy data

[a]NMSA (National Meteorological Services Agency of Ethiopia), Assessment of drought in Ethiopia, in: Meteorological Research Report Series, Addis Ababa, 1996. The agency reported severe famine that forced a change of food habits.

[b]See Degefu, W., 'Some aspects of meteorological droughts in Ethiopia,' in: M.H. Glantz (Ed.), *Drought and Hunger in Africa*, Cambridge, 1987; R. Pankhurst, The History of Famine and Epidemics in Ethiopia Prior to the Twentieth Century, London, 1985; R. Pankhurst, *The History of Famine and Epidemics in Ethiopia Prior to the Twentieth Century*, London, 1985.

[c]Pankhurst, R., *The History of Famine and Epidemics in Ethiopia.*
[d]Schove, D. J., 'African droughts and the spectrum of time,' in Dalby, D., Church R.J.H. and Bezzaz F. (Eds), ***Drought in Africa*** *2* (London, 1977); Herring, R. S., 'Hydrology and chronology: the Rodah nilometer as an aid in dating interlacustrine history,' in: F.B. Webster (Ed.), *Chronology, Migration and Drought in Interlacustrine Africa* (London, 1979).
[e]Schove, D. J., 'African droughts and the spectrum of time.'
[f]NMSA (National Meteorological Services Agency of Ethiopia), Assessment of drought in Ethiopia.
[g]Ibid.
[h]Pankhurst, R., *The History of Famine and Epidemics in Ethiopia.*
[i]NMSA (National Meteorological Services Agency of Ethiopia), Assessment of drought in Ethiopia.
[j]Ibid.
[k]Ibid.
[l]NMSA (National Meteorological Services Agency of Ethiopia), Assessment of drought in Ethiopia.
[m]Schove, D. J., 'African droughts, and the spectrum of time.'
[n]Ibid.
[o]The Borana were unhappy with the reformation brought by the *gada* leader.
[p]Pankhurst, R., *The History of Famine and Epidemics in Ethiopia.*
[q]Ibid.
[r]Herring, R. S., 'Hydrology and chronology: the Rodah nilometer as an aid in dating interlacustrine history,' in: F.B. Webster (Ed.), *Chronology, Migration and Drought in Interlacustrine Africa* (London, 1979).
[s]R. Pankhurst, *The History of Famine and Epidemics in Ethiopia.*
[t]Schove, D. J., 'African droughts and the spectrum of time'; Herring, R. S., 'Hydrology and chronology: the Rodah nilometer as an aid in dating interlacustrine history,' in Webster, F.B. (Ed.), *Chronology, Migration and Drought in Interlacustrine Africa* (London, 1979).
[u]Pankhurst, R., *The History of Famine and Epidemics in Ethiopia.*
[v]The strengthening of *gada* institution was the return of the *daacci* of the *Darara* of Mada Boru Dadoye when *gada* institutions were made more robust.
[w]Wood, C. A., 'A preliminary chronology of Ethiopian droughts,' in D. Dalby, D., Church, R.J.H. and Bezzaz, F. (Eds), Drought in Africa, (London, 1977).
[x]Herring, R. S., 'Hydrology and chronology: the Rodah nilometer.'
[y]Schove, D. J., 'African droughts and the spectrum of time.'
[z]Degefu, W., 'Some aspects of meteorological droughts in Ethiopia,' R. Pankhurst, *The History of Famine.*
[aa]Håkanssen, N. T., 'Cattle, climate and caravans: The dynamics of pastoralism, trade, and migration in 19th-century East Africa,' in *Migration in Africa* (London, Routledge, 2022 pp. 95–111).
[bb]The conflict was over transfer of power from the *gada* in power to the next creating social instability.
[cc]Dale, I. R., 'Is East Africa drying up?', *The East African Agricultural Journal*, 17 (3) (1952): 116.
[dd]NMSA (National Meteorological Services Agency of Ethiopia), Assessment of drought in Ethiopia.
[ee]Hastenrath, S., 'Variations of East African climate during the past two centuries,' *Climate Change* 50 (2001): 209–217.
[ff]Ssemmanda, I., Gelorini, V. and Verschuren, D., 'Sensitivity of East African savannah vegetation to historical moisture-balance variation,' *Climate Past* 10 (2014): 2067–2080.
[gg]C.A. Wood, A preliminary chronology of Ethiopian droughts.
[hh]Conway, D., 'Extreme rainfall events and lake level changes in East Africa: Recent events and historical precedents,' in Odada, E. O. and Olago, D. O. (Eds.), *The East African lakes: Limnology: paleolimnology and biodiversity*, pp. 62–92, Kluwer, Academic, The Netherlands.

The sixteenth century was such a period of crisis. From oral historical sources, the *gada* of Dagale Yaya (1536–1544)—the earliest time recalled—experienced megadroughts. The famine reduced human and livestock populations. This event is also reported in the Ethiopian Historical Meteorological Service publication, and the period coincided with decreasing flows of the Blue Nile, which resulted in a major famine that killed many people in Egypt, spreading into East and Central Africa.[57]

Oral historians recall further periods of megadroughts and famine during the *gada* of Asosa Titile (1544–1552), Boru Luku (1552–1560), and Goba Alla (1689–1697). These regional droughts lasted for up to seven years and devastated pastoral economies. The *gada* of Dawe Gobbo (1697–1705) was also associated with droughts and famine, with oral sources reporting that the crisis occurred 'everywhere.' Other periods of epic droughts occurred during the *gada* of Mada Boru (1752–1760), Wayu Raale (1784–1792), Saaqo Dadacha (1808–1816), and Sokhore Anna (1824–1832).

According to oral sources, epic droughts drastically reduced human and livestock demography, leaving the land 'empty.' Davis[58] reports that the famine of 1876 would have caused hundreds of thousands of deaths. The Borana oral historians refer to that drought as a 'century's drought' (*adaalanin ittii gaalte*—returning in a 100-year cycle) due to its severity that was comparable only with the *gada* of Boru Lukku (1552–1560), when similar conditions existed. Water sources (including rivers) dried up during a century's drought.

In the history of famine in Ethiopia, Richard Pankhurst[59] describes extended periods of disasters that devasted food sources in the region, resulting in massive losses of grain production and the collapse of human and stock populations. The causes were both natural and socio-politically induced. These events affected the highlands and spread across the lowlands, predominantly inhabited by nomadic societies (the Oromo and Somalis, among others). These famines had far-reaching and long-term social and economic consequences across the region. In opposition, there were cycles of wet years followed by locust plagues.

[57] Caviedes, C. N., *El Niño in history: Storming through ages* (Florida FL: University Press of Florida, 2001).

[58] Davis, M., *Late Victorian holocausts: El Niño famines and the making of the third world* (London: Verso, 2002, 284).

[59] Pankhurst, R., *The history of famine and epidemics in Ethiopia prior to the twentieth century* (Addis Ababa: Commission of Relief and Rehabilitation, 1990).

Wet-year Cycles and Locust Plagues

Borana social memory reports periods of excessive rainfall that were also considered disasters. During the *gada* of Bidhuu Dhoqqe (1568–1576), heavy rains fell continuously for eight straight years, which collapsed and destroyed the ancient *Tula* wells. The hooves of stock fell off due to rot, while swarms of mosquitoes caused a malaria epidemic among the population. During the *gada* of Ungule Lakhe (1800–1808), extreme rainfall events that fell over a noticeably brief period broke an extended drought that killed 'all trees.' The floods killed people and stocks and destroyed hydraulic infrastructure such as wells. This was the rain predicted by a local prophet called Sooga, referred to as *gaaf gaana sooga* (the time of the long rains of Sooga).

Wet years were followed by outbreaks of locusts, reported in the contemporary chronicles of Ethiopia.[60] The oral calendar reports that locust plague episodes aligned with regional famines. The periods of excessive rainfall followed by plagues of locusts occurred during the *gada* of Babbo Hooro (1616–1624) and Babo Sibu (1624–1632). The locust plagues continued into the *gada* of Indale Doyo (1632–1640), causing extended periods of famine. Further famines associated with plagues of locusts were reported during the *gada* of Guyo Gedo (1744–1752). The combined locust famine and epidemics during the *gada* of Gobba Alla (1689–1697) decimated the cattle population and broke the social structure, which was aggravated by external and internal conflicts (Fig. 3.1).

External and Internal Conflicts

External conflicts occurred with neighboring pastoralist groups—mainly the Somalis and other Oromo clans. These were significant sources of perturbations of the Borana pastoral economy during the seventeenth and eighteenth centuries. The earliest significant conflicts that dislocated the Borana from their extensive territories occurred during the *gada* of Wayu Reele (1672–1680) and in the eighteenth century during the *gada* of Dadacha Oda Morowa (1760–1768). The oral sources are not specific about the groups with whom the Borana were at war during that time. Both severe drought and further wars followed this period during the *gada* of Bule Dadacha (1776–1784), but again, the oral sources are vague

[60] Degefu, W. Degefu, 'Some aspects of meteorological droughts in Ethiopia.'

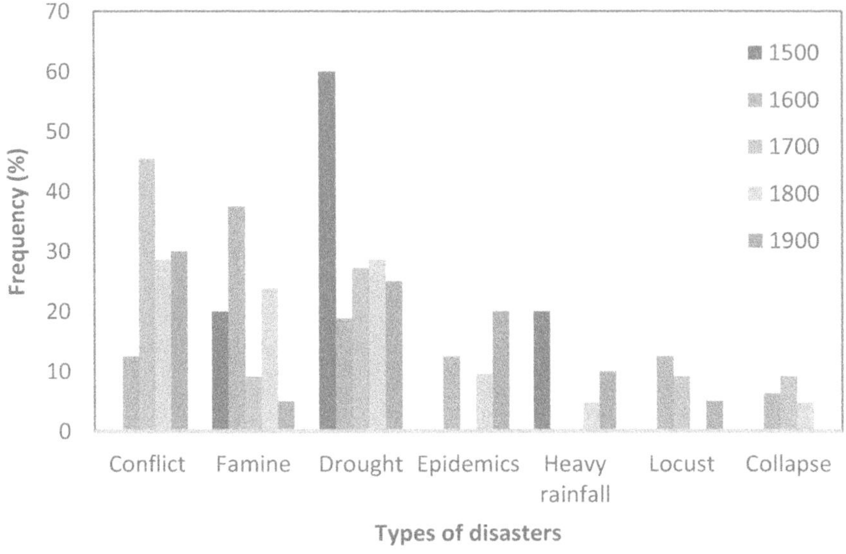

Fig. 3.1 Frequencies of disasters reconstructed using the Borana *gada* timeline

about the events. What is certain is that the society was displaced from eastern grazing lands during the second half of that century.

Internal conflicts sometimes occurred over power transfer between the generation classes when there was more than one contestant. Nonetheless, under no circumstances would delays in the power transfer change the duration of the next *gada*, which was permanently restricted to the time remaining in the specified eight-year period. This happens when internal political upheavals are associated with weakening the *gada* institution.

Oral sources recall that due to internal political competition during the *gada* of Walle Waccuu (1712–1720), the assembly did not follow rituals that were compulsory daily activities. The *gada* leader was isolated, and in protest, he left the ritual encampment and escaped into a remote area. As a rule, no ritual activities may take place in the absence of the *abba gada*. Consequently, the *gada* council became dysfunctional. The oral sources recall the period when the social system broke down. They report that the society suffered a pandemic of scabies that was viewed as a curse placed on them by the *gada* leader.[61] Further internal conflicts—also associated with competition over leadership—occurred during the *gada* of Doyo Jillo (1856–1864), Haro Adi (1864–1872), and Dida Bitata Mamo (1872–1880).

Such continual internal conflicts weakened governance, making society vulnerable to external attacks. These disasters, and those associated with droughts, triggered large-scale displacements of communities. The number of the movements reported were grazing migrations that appear to have occurred as adaptive responses to internal and external pressures. Internal decisions may have included moving the populations to regions with more reliable water and grazing resources.[62] Such internal population movements occurred during the *gada* of Abayi Babo (1656–1664) when the Borana ritual leaders crossed from Liban in the east to the central region of the ancient *Tula* wells. More far-reaching disasters were those associated with pandemics, which caused the collapse of human and animal populations.

Pandemics and Population Collapse

Periods of human and livestock epidemics were often accompanied by economic collapse. The Borana use a general term to refer to pandemics among people—*barra ch'iisa* (literally, 'periods of sleep')—caused by malaria, cholera, and smallpox. Among stock, the pandemics were bovine pleuro-pneumonia and the rinderpest epizootic. A cattle epidemic was first reported during the *gada* of Gobba Alla (1689–1697). According to oral tradition, human and stock pandemics came in multiples, causing social structures to break down and—combined with famine—resulting in the collapse of human demographics.

The disasters that are imprinted on social memories of the Borana to the present day are those of the late nineteenth century. Smallpox and cholera damaged human demography; however, the rinderpest pandemic that wiped out ungulates (including cattle) lingers as the strongest memory. In the Borana region, oral tradition suggests that only 1000 souls survived the famine that followed the mass deaths of cattle.[63] The period

[61] Interview with Godana Ajaaa by the author in 1976 in Sololo, Kenya.

[62] Tiki, W., and Oba, G. 'Righting the Wrongs: Contesting Water Property Rights in Southern Ethiopia,' *Human ecology* 45 (6) (2017): 723–734.

[63] Interview with Borbor Bulle, in 1996 in Borana, Ethiopia.

is historically remembered *as Ciina tiite guraacha*—the termination period of blue flies—when flesh-eating blue flies covered cattle carcasses.[64]

The rinderpest epizootic was widely reported across Africa.[65] Bender[66] describes the period as an 'ecological collapse,' Brown[67] describes the event as 'exceptionally erosive…when… pernicious diseases decimated the humans, and cattle died in huge numbers.' According to Helge Kjeskshus,[68] the period left historical memories of ecological crises, demographic collapses, social disintegration, and collapsed economies that had no parallels in recent recorded human history. In Ethiopia, an estimated 90 percent of cattle died. In the region of Wabbe Shebele, Captain Swayne[69] reports a famine that struck the Somali population. Drake-Brockman says that people survived on plant leaves and roots during the same period. The rinderpest epizootic was followed by smallpox epidemics (1889–1890) and cholera (1889–1892).[70] These cyclical disaster events became benchmarks for the Borana people to develop a sense of *daacci*.

The daacci *Cycles*

According to Borana's oral tradition, history repeats itself, with cycles of disasters having a recurring influence on adaptive strategies. For example, Spinage[71] reports repetitions of earlier disasters during the recent past, with similar events being predicted in the future. This repetition of histori-

[64] Tiki, W., and Oba, G., '"Ciinna"—the Borana Oromo narration of the 1890s Great Rinderpest epizootic.'

[65] Waller R., '" Emutai": Crisis and response in Maasailand 1883–1902,' in Johnson, D. and Anderson, D. M. (Eds.) *The ecology of survival: Case studies from northeastern Africa* (Boulder CO: Westview Press, 1988).

[66] Shetler, B. J., Imaging *Serengeti: A history of landscape memory in Tanzania from earliest times to the present* (Athens: Ohio University Press, 2007).

[67] Brown, *Where Giants Trod: The saga of Kenya's desert lake* (London: Quiller Press, 1989).

[68] Kjekshus, H., *Ecology, control and economic development in East African history: The case of Tanganyika 1850–1950* (London: James Currey, 1996).

[69] Swayne, Harald George Carlos. *Seventeen Trips Through Somáliland: A Record of Exploration & Big Game Shooting, 1885–1893: Being the Narrative of Several Journeys in the Hinterland of the Somáli Coast Protectorate, Dating from the Beginning of Its Administration by Great Britain Until the Present Time; with Descriptive Notes on the Wild Fauna of the Country* (London: Rowland Ward, 1895, 188).

[70] Hartwig, G. W., 'Demographic considerations in East Africa during the late nineteenth century,' *International Journal of African Historical Studies* 4 (1979): 653–672.

[71] Spinage, C. A., *African ecology-benchmarks and historical perspectives* (Berlin: Spring-Verlag, 2012, 63).

cal events forms the Borana philosophy of *daacci*. The concept should always be used in context since it augurs well or poorly for everyone: families have *daacci* of misfortunes, and among the *gada* generation classes, *daacci* of the same class returns when they take power. That fateful *daacci* will not pass—returns are inevitable.[72] The knowledge of *daacci* is provided by local time experts (*ayaantu*) who associate disaster crises with events such as changes in leadership and natural events such as eclipses of the sun. The returns of *daacci* are reported in terms of *gogessa* generation classes and the *maqbas* naming cycles.

We consider the *gogessa* cycles first. The droughts during the *gada* of Liban Jaldesa (1888–1896), Bulle Dabasa (1929–1936), and Goba Bulle (1968–1976) returned to the same *gogessa*. The drought of the *gada* of Goba Bulle, which lasted from 1973 to 1975, was the most severe among recent droughts. This was a return of the famine of the *gada* of Goba Alla (1689–1697) when all animals with 'open hooves' died.[73] There were droughts and famine during the *gogessa* of Sokhore Anna (1825–1833) and Madha Boru (1832–1840). The *daacci* of famine returned during the *gada* of Haro Adi (1868–1876) of the same *gogessa* (Table 3.1). The famine that left memories of great suffering was that of the rinderpest epidemic of 1891,[74] which lasted during the last four years of the *gada* of Liban Jaldesa (1888–1896) and was broken at the beginning of the *gada* of Adi Doyo (1896–1904). The *daacci* returned during the *gada* of Liban Kusse (1913–1921) when cattle died of contagious bovine pleuropneumonia (CBPP) (see Table 10.1).

We now summarize the types of disasters corresponding with the *maqbas* cycles of the *gada* timeline (Fig. 3.2). Six of the seven *maqbas* classes experienced inter-ethnic conflicts, two experienced droughts, and five *maqbas* experienced famine. The *Darara maqbas* experienced the lowest number of disasters. The return of the *maqbas* was associated with other cyclical political and climatic events and epizootics (see Table 3.1).

The Borana used various cosmic signs to predict coming droughts and famine, such as eclipses of the sun. During the famine of the *gada* of Liban Jaldesa (1888–1896), the solar eclipse occurred two years before he

[72] Waktole Tiki & Gufu Oba 'Ciinna—the Borana Oromo narration of the 1890s Great Rinderpest epizootic in Northeastern Africa,' *Journal of Eastern African Studies*, 3:3 (2009), 479–508. 10.1080/17531050903273750.

[73] Interview with Oba Sarite Kura, Sololo, 1993.

[74] Waktole Tiki & Oba, G. 'Ciinna—the Borana Oromo narration of the 1890s Great Rinderpest epizootic.'

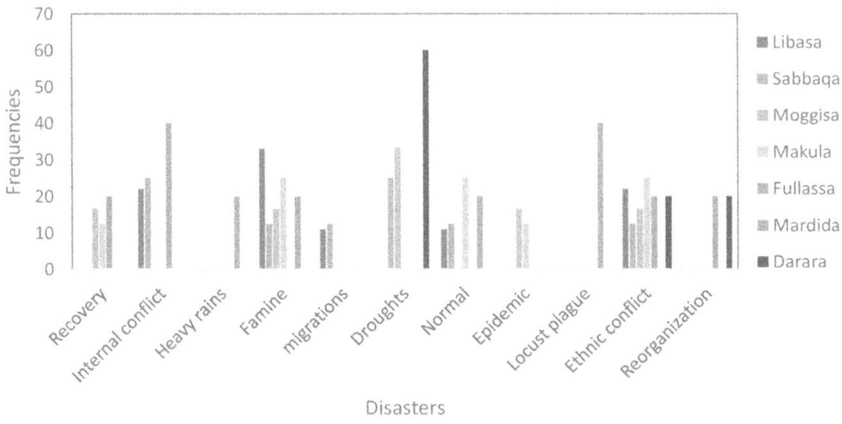

Fig. 3.2 Frequencies of types of disasters according to the maqbas cycles in the Borana event historical calendar

handed power over to Adi Doyo (1896–1904). The population suffered severe droughts during those two years. Another solar eclipse occurred during the *gada* of Bulle Dabassa (1929–1937) when a severe drought was again reported. The next solar eclipse occurred during the *gada* of Goba Bulle (1968–1976), coinciding with the infamous Sahelian disaster in 1973.[75]

From the Borana perspective, the *daacci* has been notoriously accurate in the past in terms of predicting the future. For example, the *sabbaqa* of Walle Wacuu (1713–1721) returned with Boru Mad'a (1993–2001) after 280 years. These were both periods of famine. The *Darara* of Dawe Gobbo (1697–1705) returned with Jillo Aga Adhi (1977–1985), both being periods of peace and recovery. The famine during the *gada* of Dida Bitata Mammo (1873–1881) and the return of the *Libasa* of Boru Luku (1593–1601) was associated with a 'worldwide' famine. The *Libasa* of Bulle Dabasa (1929–1937) is also associated with severe droughts which killed livestock and people.

How might we then explain the persistence of Borana pastoralism during these centuries? One fact is clear: disasters were followed by periods of recovery. The time intervals varied from years to decades. This shows

[75] Sinclair, A. R. E., and J. M. Fryxell. "The Sahel of Africa: Ecology of a disaster." *Canadian Journal of Zoology* 63.5 (1985): 987–994.

Borana pastoralism has remained resilient despite prolonged and recurrent disasters over the previous centuries.

The Restoration of Pastoral Resilience

Social memory narratives of the Borana report recoveries after disasters. Periods of recovery might be examined from two perspectives. Primarily—considering that strong institutions guided the Borana society—it was necessary to consolidate and restore the social institutions of the *gada* that could then plan to repair damage resulting from disasters. The Borana used an interesting metaphor to support the *gada* rituals during disasters when cattle pastoralism had collapsed. According to oral historians, *gada* activities became more symbolic when fruits of Adam's apple shrub (*hiidi*) or *Solanum* species served as 'cows' and were 'slaughtered' to fulfill the necessary *gada* rituals.[76] Second, more fundamentally, recovery implied the revival of social institutions for reorganizing the society—seeking out lost members of the clans and developing mechanisms for restocking the population.

The first step was to strengthen the *gada* institution to buffer the Borana clans, and the second step, as recovery progressed, was to reorganize the families into cooperate groups which, in the long term, revived the functions of the clan institution of *buusa gonoofa* (see Chap. 8). This included repairing hydraulic infrastructures that had fallen into disuse due to negligence during crises. The *gada* of Morowa Abay (1680–1688) was one such period when the *Tula* wells were rehabilitated, and other periods of recovery occurred during the *gada* of Sora Dadacha (1729–1737) and Boru Madha (1792–1800) (including strengthening of the *gada*), and finally, the *gada* of Adi Doyo (1896–1904) just before the Borana were incorporated into colonial empires. During the last four years of the *gada* of Adi Doyo, the Borana pastoral economy was on the path to recovery after the Rinderpest disaster. The Borana refer to the period as *jiireena gada* Adi Doyo (the prosperity of the *gada* of Adi Doyo).

The late nineteenth century was also when European travelers presented eyewitness accounts of the revival of pastoralism.[77] Arthur Donald

[76] Interview with Borbor Bulle by Gufu Oba in Nagelle Borana, Ethiopia, in 1997.

[77] Lamphear, John. "Aspects of 'becoming Turkana.'" *Being Maasai: Ethnicity and Identity in East Africa. London, James Currey* (1993): 87–104.

Smith,[78] the American medical doctor and explorer who crossed the Borana country seven years after the rinderpest disaster, found evidence of recovery. James McCann[79] reports that the period from 1889 to 1892 was a time of recovery in the Ethiopian highlands. That period was followed by political tragedy, namely the colonial partitioning of pastoralist territories in northeast Africa. This is the subject of Chap. 4.

[78] Smith, A. D., *Through unknown African countries: The first expedition from Somaliland to Lake Lamu* (New York: Edward Arnold, 1897); 'Dr. Donaldson Smith's journey in Somaliland,' *The Geographical Journal* 5 (1895): 124–127.

[79] McCann, J. C., *From poverty to famine in northeast Ethiopia: A rural history, 1800–1935* (Philadelphia, PA: University of Pennsylvania Press, 1987, 3).

CHAPTER 4

Colonial Trans-frontier Grazing Controls: Responses to Political and Climatic Shocks, 1908–1962

In this chapter, we gauge how pastoralists incorporated a variety of adaptive strategies to ecological and political risks in exploiting frontier resources.[1] This is significant, considering that colonial encapsulation of the frontier pastoralists occurred a decade after the catastrophic collapse of the Rinderpest pandemic in 1891 (Chap. 3). No sooner had recovery occurred that the pastoralists of the southern frontier were exposed to the combined effects of political and climatic shocks. The chapter discusses how colonial imposition of political borders had wide-ranging impacts on pastoralist adaptations in the context of climatic history. We use the Borana, a southern frontier community, as a case study to investigate their adaptive responses to the effect of international borders and frontiers between southern Ethiopia and northern Kenya in controlling trans-frontier grazing during the twentieth century. We examine the impact of the colonial government on pastoralism adaptive strategies. We consider how pastoralists adapt to political pressures when responding to trans-frontier grazing. An important question is how pastoralists sustain the resilience of their herds during periods of drought and heavy rainfall. For nomadic pastoralists split by international borders, political shocks impaired cross-border

[1]Wright D. K., 'Long-term dynamics of pastoral ecology in Northern Kenya: An old model for new resilience,' *Journal of Anthropology Archeology* 55 (2019)101068.

G. Oba, *Pastoralist Resilience to Environmental Collapse in East Africa since 1500*, https://doi.org/10.1007/978-3-031-48291-5_4

mobility.[2] Ecological conditions, the distribution of dependable resources (water and grazing), and the politics of the trans-frontier region influenced the relationships between states and communities across the frontier. Peter Rigby's[3] work concerns problems of politico-economic transformation of pastoral societies by colonialism. Further, Daniel Bates[4] identifies two processes triggered by the location of colonial borders. First, inter-ethnic relations across resource borders can integrate frontier societies to tackle environmental and social shocks (see Chap. 8). Second, within a regional political milieu, the exploitation of essential resources across ecological and social-political niches positively influenced pastoralists' adaptive strategies.[5] Invariably, however, political borders served as barriers for pastoral nomadic peoples, with disruptions of movements contravening the flexibility needed for exploiting the variable ecology and human relations.[6]

We examine these relations from the perspective of the anthropological theory of borders and frontiers, incorporating negotiations across cultural boundaries.[7] These have social, cultural, and economic dimensions.[8] On the other hand, the anthropological analysis of relations across international frontiers focuses on material and information transfers. In contrast, the implication of losing common resources critical for the survival of communities across boundaries should be discussed more. When there is political interference, reciprocal social systems may reach a breaking

[2] Markakis, J., *Resource conflict in the Horn of Africa* (London: Sage, 1998).

[3] Rigby, P., *Persistent pastoralists: Nomadic societies in transition* (London: Zed Books Ltd., 1985).

[4] Bates, D. G., 'The role of the state in peasant-nomad mutualism,' *Anthropological Quarterly* 44 (3) (1971): 109–131.

[5] Tonkin, E., 'Borderline questions: people and space in West Africa,' in Donna H. and Wilson T. M. (Eds.), *Border Approaches: Anthropological Perspectives on Frontiers* (University Press of Amer, 1994); Phiri S. H. 1984, 'National integration, rural development and frontier communities: The case of the Chewa and Ngoni astride Zambian boundaries with Malawi, and Mozambique' in Asiwaju, A. I (Ed.). *Partitioned Africans: ethnic relations across Africa's international boundaries 1884–1984* (Lagos: C. Hurst & Co., 1984).

[6] Donanan H. and Wislon, T. M., 'An anthropology of borders of frontiers,' in Donnan H. and Wilson, T. M., *Border approaches: Anthropological perspectives of frontiers* (University Press of America, 1994).

[7] Donnan H. and Wilson, T. M., *Border approaches Anthropological perspectives of frontiers* (University Press of America, 1994).

[8] Tronvoll, K., 'Borders of violence-boundaries of identity: Demarcating the Eritrean nation-state,' *Ethnic and racial studies* 22 (6) (1999): 1037–1060.

point.[9] For the inhabitants of zones that share borders and frontiers, political boundaries are inconvenient and anathema to their traditional need for mobility across the frontier in response to resource variability. For whatever reason—it undermines their traditional adaptive strategies to environmental and political shocks.[10] It fails to account for ecological relations in herder adaptations to climate variability.

The chapter is divided into four parts and sub-sections. The first part briefly describes the political ecology of contested frontiers and the effects of trans-frontier policies on traditional grazing patterns. The second part introduces the Borana, one of the frontier pastoral communities, and their adaptive responses to political and grazing variability. The third section scrutinizes climate history and its impacts on NFD grazing patterns. The fourth section discusses the findings in the context of continuity and persistence of pastoral resilience.

The Political Ecology of Contested Frontiers

The imposition of international borders separated migrating nomads between different states. Superimposed on the political control of herd movements was the variable climate that pushed land-use activities in opposite directions in response to the distribution of water and grazing resources. The periods for the discussions cover 1908–1962. We examine how pastoralists built adaptive strategies to promote the resilience of herd management by maintaining long-distance mobility. In this context, political ecology presents a robust framework through which the adaptive cycle of pastoral resilience might be analyzed.[11] The factors affecting the strength and collapse of the rural economy were numerous. What we have, in fact, are composite factors that affected the resilience of the trans-frontier pastoral system—periods of drought, famine, and epidemics—interspersed with very wet years—influenced stability (see Fig. 4.1).

We now introduce the colonial frontier conditions that resulted in conflicts, thus continuing to influence pastoralists' adaptations to climate variability. The advent of the colonial administration created political borders

[9] See Bauer, K. M., *High frontiers: Dolbo and the changing world of Himalayan pastoralists* (Colombia University Press, 2004).

[10] Chang C. and Koster, H. A., *Pastoralists at the periphery: Herders in a Capitalist World* (Tucson: The University of Arizona Press, 1994).

[11] Wright, Katy. *Community Resilience: A Critical Approach* (London, Routledge, 2021).

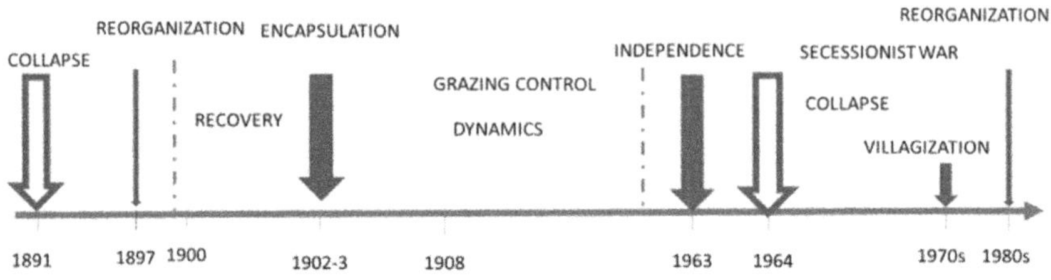

Fig. 4.1 Schematic model of resilience depicting the composite effects of various forces affecting the strength of pastoralism on the southern frontier

that separated different pastoral communities; the shift from pre-colonial dynamic land-use systems ignored how pastoralists had managed land as a communal resource in response to climate variability. Further, the seemingly random existence of pastoralist families with their herds went against the central control elements that the colonial states imposed.[12] The contradiction is apparent; the administrative control of pastoral movements created new grazing frontiers that negated pastoralists' adaptive strategies. As a result, the administration experienced difficulties controlling frontier pastoralists due to their search for scarce resources. We examine responses by the Borana pastoralists.

The Borana as Trans-frontier Pastoralists

The Borana historically was one of the dominant populations in northern Kenya and southern Ethiopia.[13] The Borana conquest by Menelik forces in 1897 was motivated by economic and political goals. Economically, the south Ethiopian highlands were settled by soldier–settlers due to their suitability for crop cultivation.[14] By the early twentieth century, all the sub-humid zones in the southern frontier with the potential for

[12] Gilliland, H. B., 'An approach to the problem of the government of nomadic peoples: With special reference to experience in Eastern British Somaliland,' *South African Geography* 29 (1) (1947: 55): 43–58.

[13] KNA/PC/NFD1/1, Christie-Mitler, D. G. to McKenzie Smith Moyale District handing over report 1949, p. 15.

[14] Oba, G., *Nomads in the shadows of empires: Protests, contests and legacies in southern Ethiopia and northern Kenya frontier* (BRILL, 2013).

agriculture had been 'colonized' by plow agriculture.[15] For the British, the importance of the Borana country was in terms of its supplies of raw materials for trade and collections of the Borana cattle breed and horses,[16] which was popular with the European settlers in the Kenyan highlands and ranches on the Laikipia Plateau.[17] The border-left primary water sources (the ancient water systems of the nine *Tula* well complexes) on the Abyssinian side of the frontier. At the same time, the wet season grazing lands lay in British territory.[18] The wells became the nodal points for Ethiopian tax collection during the dry season when livestock from within and across the frontier converged on them.[19]

Any attempt by the Ethiopians (or Italians) to deny pastoralists access to water sources across the frontier resulted in a 'loss of...stock in dry weather' on the British side.[20] Indeed, the folly of the British administration in controlling pastoral grazing movements was acknowledged— '[which] ... served to show how completely helpless ... tribesmen would be should the frontier be closed.'[21] The colonial states' attempts at control, while endeavoring to solve problems of resource scarcity, had the opposite effect. Although the 1907 trans-frontier grazing treaty facilitated movements of the Borana across the international frontier, its implementation was a bone of contention between the Ethiopian and the British and later Italian colonial administrations. These disagreements hurt the coping strategies of the Borana pastoralists.[22] While sympathetic to Borana's maltreatment at the hands of the Ethiopians and later the Italians in 1935, the Kenyan colonial administration provided no relief to the pastoralists.

[15] Markakis, J. and Ayele, *Class and revolution in Ethiopia* (Nottingham: Spokesman, 1986, 29).

[16] Huxley, E., *Whiteman's country. Lord Delamere and the making of Kenya, Vol 1 1870–1914* (London: Chatto & Windus, 1953).

[17] The Earl of Lytton, *The stolen desert: A study of Uhuru in Northeast Africa* (London: McDonald, 1966).

[18] Gwynn, C. W., 'A journey in southern Abyssinia,' *The Geographical Journal* XXXVIII (1911): 113–134.

[19] Interview with Anna Waqo Dogo, interview by the author in 1992 in Moyale Ethiopia in 1992.

[20] KNA/PC/NFD1/1/3, Northern Frontier Annual Report for 1950.

[21] Ref. no. AGR/11/3, From H. W. Wallee 1951, District Commissioner's Office Moyale, Northern Province to PC Isiolo, 12 April 1951.

[22] KNA/PC/NFD/4/3/2, H.B. Kittermaster, The NFD, Provincial Annual Report 1917–1918.

Since the Italians were not signatories to the trans-frontier grazing treaty of 1907, they did not take kindly to the Kenyan administration's demands for full access by British pastoralists to water and grazing across the frontier.[23] For example, on February 15, 1938, the Italian Major Signorini, who was the *commissario* of the Borana Province, visited Sololo (i.e., the Obbu region) and instructed that all the British Borana who wished to water in Italian territory should possess a paper pass signed by the *residente*.[24] The Kenyan colonial administration objected to this policy because the treaty's terms honored free access by the border community to the nearest wells. The Italians argued that they interpreted the policy as 'the tribe's people should be subject to territorial jurisdiction and that the Italian government was…empowered to tax them.'[25] When the Ethiopian administration returned to power six years later, conditions for the Borana across the frontier did not improve.[26] The rigidity of the Ethiopian authorities against the Borana continued to undermine their adaptive strategies to rainfall variability. We examine the impact of climatic variability and grazing patterns on Borana herd movements in NFD and across the frontier.

Impacts of Climate Variability and Grazing Patterns

Over the years, climate variability has left its marks on the region's pastoral economy and cultural history.[27] Disease epidemics and outbreaks of locusts aggravated the stresses caused by droughts.[28] We have access to impressive colonial records of climatic conditions and herd adaptive responses over five decades. We analyzed decadal climate fluctuations, grazing patterns, epidemics, and locust plagues that influenced the grazing strategies of the

[23] KNA/DC/MLE/3/4, Provincial monthly intelligence report for the month of October 1934.

[24] PC/NFD 1/1/6, Moyale District, Northern Frontier, annual Report for 1938, 6.

[25] Ibid.

[26] KNA/PC/NFD/1/1/9, Northern Province Annual Report for 1949.

[27] Andrezjewski, B. W., 'Images of drought in Somali literature,' in Lewis, I. M. (Ed.) *The Somali drought* (London: IAI, 1975, 15); Oba, G., '"Where the bulls fight, it is the grass that suffers:" impact of border administration on drought-coping strategies of the Obbu Booran during the 20th century.' *Journal of Oromo studies* 7, no. 1&2 (2000): 87–108.

[28] Mohamed, J., 'Epidemics and public health in the late colonial Somaliland,' *Northeast African Studies* 6 (1–2) (1999): 45–81.

Borana pastoralists. We present the findings chronologically over 54 years in the following sub-sections.

Adaptations to Climate Variability and Epidemics, 1908–1919

The earliest records of the impacts of climatic variability on grazing movements are from the Mandera District, which was reported to be very dry in 1908. Rainfall failure and the severe rinderpest plague caused abnormal pastoral herd mortalities.[29] The epidemics during 1907 and 1908 had 'carried away' 60 percent of the Somali cattle population.[30] Major Gwynn,[31] conducting a boundary survey, reports: 'The rinderpest had been particularly bad on the Borana highlands where high cattle mortalities occurred.' The years 1909 to 1910 were average years.

1911, 1912, and 1913 were again periods of severe droughts. The rainfall also failed in 1914, and the rinderpest was persistent, aggravating cattle mortality.[32] The Wajir District report of 1914 is quite candid on the consequences of the rainfall failure, in that high numbers of livestock had perished.[33] The need for additional water supplies became urgent in 1915. By January of that year, the Moyale District was experiencing severe food shortages following the failure of the April rainfall, which resulted in the loss of large herds.[34]

The Moyale District report of October 1915 mentions an outbreak of an unidentified virulent disease among cattle—known locally as *kurkur*—which 'wiped out a great number of cattle.' The administration prevented the pastoralists from using the foothills of the border escarpments for fear of spreading the disease. The subsequent failure of the short rains in November 1915 caused fear of the return of famine.[35]

[29] KNA/DC/MDA/4/2, Political records, Mandera NFD report by Major Gwynn 1909.

[30] KNA/DC/MLE/2/2, Moyale District, political records 1902–1929.

[31] Gwynn, C. W., A journey in southern Abyssinia, *The Geographical Journal* XXXVIII (1911): 113–134.

[32] KNA/PC/NFD/4/3/2, Wilfred Thesiger, H.B.M. Minister to Officer Commanding Moyale, 18 September 1913, Abyssinian Affairs.

[33] KNA/PC/NFD/4/1/2, Political record book, Wajir District, Northern Frontier Province.

[34] KNA/PC/NFD/4/3/2, From Vincent G. Glenday, Assistant District Commissioner Moyale to Wilfred Thesiger, H.B. M's Extraordinary and Minister Plenipotentiary Addis Ababa, 4th September 1915, Abyssinian Affairs.

[35] KNA/PC/NFD/4/3/2H.B. Kittermaster, Abyssinian Affairs, correspondence with Addis Ababa 1913–1923 of 31 December 1915, Abyssinian Affairs.

The rains returned in March 1916 with heavy storms that broke the severe dry weather and spread throughout the district. The heavy rain recorded in Moyale fell on the day that the wells had reached an all-time low. By August 1916, the rains had stopped, and natural water sources on the Kenyan side of the frontier were running low. Fortunately, continuous rainstorms occurred over seven days, as reported on December 31, 1916.[36]

During 1917, historic rainfall was experienced, with floods that filled up the Malalba depressions in the Golbo lowlands of northern Kenya (see Fig. 2.3). According to the Moyale district annual report, the exceptionally wet rainy season from April to June 1917 'made available large areas of grazing.' The report continues to add, 'The lag Urbur from Abyssinia ran so high as to fill the Malalba depression, and this provided water for thousands of head of stock of Borana and Ajuran for several months. It is the [second] time for 15 years that this depression has filled.' The heavy floods, the water in the natural depressions, and the flooding of the Uaso Nyiro River had 'caused the country all around the Lorian swamp to be inundated for many miles.'[37] This resulted in plentiful grazing throughout the NFD, thus making any annual dry season movements for watering across the frontier unnecessary.[38]

The long April–May rains in 1918, which had been anticipated with hope, were patchy. Indeed, another regional drought again created hardship for pastoralists and their livestock. In the Buna region, the livestock depended on the wells for watering for six continuous months. During this time, herders were forced to walk their cattle long distances to reach the water due to lack of grazing. The herders watered their cattle at intervals of four days, resulting in extreme thirst for the animals. The population was forced to move outside the areas allocated to them for grazing.[39]

By July 1918, drought conditions were widespread. The short rains from October to December failed, resulting in severe water scarcity and grazing.[40] In 1919, the long (March–June) and the short (November–December) rainy seasons were failures. Whereas the Moyale District

[36] Ibid.

[37] KNA/PC/NFD/4/3/2, H.B. Kittermaster, The NFD, Provincial Annual Report 1917–1918.

[38] KNA/PC/NFD/4/1/2, Political record book Wajir NFP for 1917.

[39] KNA/PC/NFD/4/3/2., H.B. Kittermaster, Abyssinian Affairs, correspondence with Addis Ababa 1913–1923 of 31 December 1915, Abyssinian Affairs.

[40] KNA/PC/NFD/4/3/2., H.B. Kittermaster, Abyssinian Affairs, correspondence with Addis Ababa 1913–1923 of 31 December 1915, Abyssinian Affairs.

Commissioner report mentions some rains breaking along the frontier,[41] the rainfall conditions throughout NFD were dire. The provincial annual report suggests that 'Grazing everywhere was scarce.'[42] This was a regional drought influenced by La Ninña that affected the whole region of the Horn of Africa.[43] The climate situation had become so severe that it was beyond the administration's capacity to help the nomads.

Impacts of and Adaptations to Droughts, 1920–1929

There were slight improvements in the rains and the availability of grazing and water in the 1920s. The heavy rains on the Kenyan side of the frontier attracted pastoralists from across the border. This was also when the administration relaxed its control, suggesting that there was 'room for many more camels in the district.'[44] Considering the unpredictable rainfall in the NFD, shifting between extremes, this was a rarity. In the handing-over report, Plownan informed the incoming District Commissioner, Mr. Wadington, about the grazing situation along the Daua River, where the short rains of 1920 had been a failure. He hoped that the previous year's remaining grazing reserves would suffice as a fallback resource.[45]

Another widespread drought affected the NFD in 1921, with further severe repercussions for the livestock economy. The scarcity of grazing and water left the administration with little choice but to leave the pastoralists to their own devices. Indeed, the pastoralists in the worst-hit districts, such as Wajir, migrated to the Moyale District, where conditions were less severe.[46] During the year, more significant populations of frontier nomads watered across the frontier or migrated to the Ethiopian side.[47] The year 1922 was generally dry, with some wells running low, although grazing conditions along the border remained moderately favorable.[48]

[41] KNA/PC/NFD/4/3/3, From District Commissioner's Office, Moyale to the Chief Secretary Nairobi, 5th April 1919, Abyssinian Affairs.

[42] KNA/PC/NFD/4/3/2, H.B. Kittermaster, The NFD, Provincial Annual Report 1918–1919.

[43] Caviedes, C. N., *El Niño in history: Storming through ages* (Florida: University Press of Florida, 2001, 193).

[44] KNA/DC/MDA/1/1, Annual Report of the Garre District for 1920–1921.

[45] KNA/PC/NFD1/1/, Handing over Provincial report of 1920 from C. H. Plowlan to E.J. WaDington.

[46] KNA/PC/NFD/4/1/2, G. Waddington Provincial Annual Report NFD 1920–1921.

[47] KNA/DC/MLE/3/7, Moyale District Monthly Report for April 1939.

[48] KNADC/MLE/3/1, Moyale Intelligence report for 1922.

During 1923, pastoralist herds suffered from a scarcity of grazing resources and outbreaks of bovine pleuro-pneumonia. The April rains came in storms that varied from heavy to slight showers.[49] Favorable rains followed these in 1924 and 1925. For pastoralists of the NFD, grazing and water tended to be at opposite ends; where water was available, grazing was scarce, and in areas that lacked water, grazing was abundant.

The late 1920s saw a return of regional drought when comparable stresses were reported throughout northeast Africa. There were two reasons why the colonial administration had attempted to solve local drought stresses by moving pastoralists around—either by relaxing grazing controls or allowing them to migrate during severe droughts such as those of the 1920s. Firstly, high livestock mortalities demanded government intervention. Secondly, there was a perception that extreme overgrazing levels that usually followed periods of drought required measures for environmental rehabilitation once the drought conditions had broken. Patterns of rainfall in the province had remained highly unpredictable, with years of failed rainfall followed by years of heavy rains. This was the case in 1925 and 1926 when the provincial administration reported heavy rain and plentiful grazing for stock.[50] However, this was not sustained into 1927.

The month of January 1927 opened with an extended period of dryness. The Moyale District received: 'insufficient quantities to materially affect the grazing and water.' The long rains also failed when they had been expected.[51] By May, conditions for grazing and water were worsening. The reaction of the British pastoralists was to cross the border into Ethiopia.[52] According to the Wajir District Report, the situation was a famine, with an estimated death of 90 percent of the stock.[53]

By June 1928, a shortage of water and grazing was reported everywhere.[54] Between September and December, the weather conditions changed for the worse. The lack of rain on both sides of the frontier meant that all traditional pastoral grazing and watering movements were disrupted. The scarcity of water and grazing on the British border forced the Borana again to cross to the Ethiopian side.[55] By October 1928, the

[49] KNA/MLE/3/1, Moyale Intelligence report for 1923.
[50] KNA/PC/NFD/1/1/3, Northern Frontier Province Annual Report for 1926.
[51] KNA/DC/MLE/3/1, Moyale Intelligence report for 1928.
[52] Ibid.
[53] KNA/PC/NFD/1/1/3, Northern Frontier Province Annual Report for 1928.
[54] KNA/DC/MLE/3/1, Gerald Reece, Moyale Intelligence report for June 1928.
[55] KNA/PC/NFD/5/2/1, NFD Intelligente report for August-September 1928.

livestock conditions throughout the NFD had deteriorated. The livestock that had migrated to the Lorian swamp area 'is now dead.' By November of that year, there were glimpses of hope, with showers falling in some parts of the Moyale District, encouraging the Borana and other pastoralists to return to the British side of the border.[56] The rains in November 1928 were so heavy that motor roads became impassable.[57] Then another natural disaster struck, with swarms of locusts arriving and devouring vegetation.[58] The provincial annual report 1928 claims that 'hoppers hatched out in enormous numbers causing much damage to the grazing.'[59] Between August and October, conditions were so severe that 'average losses of cattle, sheep, and goats probably amount to between 80 to 90 percent.'[60]

The rainfall in 1929 came in isolated showers. For example, there were isolated showers in different parts of the province during February and March, but generally, there were drought conditions elsewhere. Again, the long rains in April did not materialize, resulting in water scarcity. There was a brief period of rainfall in May when heavy rains were reported in the Daua River area. In the Moyale District, 'for the third year in succession,' the rains from June through September failed, forcing the Borana families to cross the border again onto the Ethiopian side of the frontier. The Borana of Debel had abandoned the wells and migrated into the narrow border area that had received the rains.[61]

It was distant heavy rains in the Isiolo District that flooded the Uaso Nyiro River that 'saved the Borana living in the vicinity of the Lorian swamp from the severest losses in stock from famine in 1929.' Again, grazing scarcity was aggravated by locust swarms that spread throughout the province.[62] Locally, the drought was called the drought of locusts (*oola awaanisa*).[63] The Moyale District report claims that 'locusts consumed most of the grazing and vegetation in the Daua River valley,' causing

[56] KNA/PC/NFD/5/2/1, Intelligence report of Wajir 1927–1930.

[57] Ibid.

[58] Ibid.

[59] KNA/PC/NFD/1/1/3, Northern Frontier Province, Annual Report for 1928.

[60] KNA/PC/NFD1/5/1, Annual Report, Wajir District for 1928.

[61] KNA/PC/NFD/5/2/1, Intelligence report for Wajir District 1927–1930.

[62] KNA/DC/MDA/1/1, Gerald Reece, Gurreh District Intelligence report for September 1928, Letter ref. No. 258/1/5/528.

[63] Interview with Oba Sarite Kura in 1992, Sololo, Kenya.

widespread starvation among pastoral herds.[64] The 1930s and 1940s reflected similar patterns of climatic variability compared to previous decades.

Famines and Trans-frontier Grazing, 1930–the 1940s

The decade of the 1930s brought mixed conditions throughout the province. Despite its characteristic spatial and temporal variability, the rainfall was widespread in various districts. In Moyale, the early arrival of the rains improved stock conditions, with the pastoralists assuming their traditional wet and dry season patterns of movement, alternating between crossing the frontier and moving into the Golbo lowlands of the NFD. Locust swarms caused considerable damage to crops and grazing around Moyale and Mandera District, threatening livestock with starvation, but conditions were suitable for the herders elsewhere.[65]

Climate conditions in 1932 did not improve from the stress of the previous years. By the year-end, the province was again experiencing shortages of grazing and water. The administration began discussions on the possibility of sinking boreholes.[66] The crisis of water and grazing scarcities experienced on both sides of the frontier undermined the nomads' adaptive strategies.[67] The short rains throughout the region—including the traditional drought escape frontier with Ethiopia—were either inadequate or considered a failure.[68]

The 1933 report of the Mega British Consulate for southern Ethiopia describes widespread drought, while the provincial office for the NFD reported that the deficit was 'severe everywhere.'[69] The Uaso Nyiro River dried up for nearly four months, indicating the severity of the drought.[70]

[64] KNA/DC/MLE/2/2, Moyale District, Political records 1902–1929.

[65] KNA/PC/NFD/4/1/2, Political record book for Wajir district, 1931.

[66] KNA/DC/MDA/1/8, Northern Frontier Province Annual Report for 1931.

[67] KNA/DC/MLE/3/3, Moyale District, NFP, Monthly Intelligence report for the month of January 1932.

[68] KNA/PC/NFD/4/3/7, Mega Consulate Intelligence Report for the months of November and December 1932, Abyssinian Affairs, Mega Reports 1932–1935.

[69] KNA/PC/NFD/4/3/7, From Gerald Reece to Major A.T. Miles, Mega Consulate, southern Abyssinia, handling over report April 1933, Abyssinian Affairs 1927–1934, Mega Reports 1932–1935, Addis Ababa Reports 1944–1946.

[70] KNA/PC/NFD/4/3/7, Mega Intelligence Report for the Quarter ending September 1933, Abyssinian Affairs, Mega Reports 1932–1935.

By March 1933, there were still no rains throughout the province.[71] A crisis developed on the Kenyan side of the frontier as the Ethiopians blocked access to water and grazing on their side. The drought stress and rinderpest resulted in mass cattle mortality among the Borana herds.[72] The harsh climatic conditions were reversed by 1934 when the province and the Ethiopian side of the frontier recorded well-distributed rains.[73] This was a rare event, with administrators reporting the sufficiency of grazing and water resources.[74]

The favorable pasture and grazing conditions during the second quarter of 1935 made it unnecessary for the British Borana to cross the frontier.[75] Nevertheless, when the Ethiopian–Italian war broke out in 1935, access to grazing and water mattered most to the nomads. The favorable rainfall conditions in 1936 maintained the grazing movements of the British nomads within the province for a while.[76] However, by 1937, the pastoralists shifted to the side better endowed with resources, regardless of the political crisis.[77] By 1939, the nomads who had crossed the frontier were returning to British territory.[78]

The 1940s were transitional years in the administration of the southern Ethiopia–northern Kenya frontier. After defeating the Italians, the British—temporarily occupying the Ethiopian side of the border—handed the administration over to the Ethiopians. The long rains in April 1940 had broken the previous drought conditions.[79] The British administration was at loggerheads with the Ethiopians over trans-frontier watering rights. Nevertheless, this turn of events did not last, as the following year (1941)

[71] KNA/PC/NFD/4/3/7, Mega Intelligence Report for the Quarter ending September 1933, Abyssinian Affairs, Mega Reports 1932–1935.

[72] KNA/DC/ISO/2/1/2, Northern Frontier Province Annual Report.

[73] KNA/PC/NFD/4/3/7, Mega Consulate Intelligence Report for the last Quarter 1934, Abyssinian Affairs, Mega Reports 1927–1934.

[74] KNA/DC/MLE/3/4, Provincial monthly intelligence report for the month of October 1934.

[75] KNA/PC/NFD/4/3/7, Mega Consulate Intelligence report for the second quarter 1935.

[76] KNA/PC/NFD/4/1/2, Moyale District NFD Resume of happenings in 1936 (for political records).

[77] KNA/DC/MDA/1/2, Annual Report to District Commissioner, Moyale on Mandera sub-District for year 1937.

[78] KNA/DC/MLE/3/7, Moyale district monthly report for April 1939.

[79] Ref. No. Publ. 24/2/2, KNA/DC/MLE/3/7, Moyale Northern Frontier-Intelligence Report for April 1940.

was generally drier.[80] Indeed, the dry climatic conditions were widespread throughout northeast Africa. The situation was again aggravated by locusts that produced great suffering.[81]

In 1942, the Moyale District recorded the lowest rainfall since 1916.[82] This resulted in high livestock deaths and widespread famine throughout Ethiopia and Kenya. In the Mandera District alone, 1400 people were in famine relief camps.[83] The Marsabit District report states, 'This year [1942] has been an evil one for everybody because of severe drought and shortage of water and grazing everywhere.'[84] The famine induced refugees from the Ethiopian side of the frontier, and these impoverished nomads created another problem for the administration.[85]

The problem with the 1943 showers of rain was their patchiness.[86] Failure of the long rains and the late arrival of the short rains 'did little help to grazing.'[87] For the administration, all their attempts to control and regulate the movements of the nomads broke down during droughts, as was the case in 1943.[88] This inevitable situation forced the British Borana to cross once again into Ethiopia.[89]

In 1944, the rainfall was well-distributed and above average.[90] Nevertheless, as usual, the problems of northern Kenya come in doubles—when the rains improved, disease epidemics and pests such as locust plague devastated the grazing lands—creating an artificial drought. The subsequent famine caused 'untold losses in cattle.'[91] In 1945, although the

[80] KNA/PC/NFD1/6/4, Moyale District Northern Frontier Annual Report for 1941.

[81] Ali, S. M., *Huddur and the history of Southern Somalia* (Cairo: Nahda Bookshop Publisher, 2005, xi).

[82] KNA/PC/NFD1/6/4, Moyale District Northern Frontier Annual Report for 1942.

[83] KNA/PC/NFD1/1/8, Gerald Reece, Northern Frontier District of Kenya Annual Report for 1943.

[84] KNA/PC/NFD/6/4, From R. Tatton Brown DC Moyale to PC Isiolo, AGR/3/40, 7 December 1943.

[85] KNA/DC/MBT/, J.K.R. Thorp 1943, Marsabit, Northern Frontier handing over report.

[86] KNA/PC/NFD/6/4, Moyale District Northern Frontier Annual Report, AGRR/11/3, August, 943.

[87] Ibid.

[88] Ibid: 5.

[89] KNA/NFD1/1/8, Gerald Reece, Northern Frontier District of Kenya Annual Report for 1943.

[90] KNA/PC/NFD1/6/4, Moyale District northern Frontier Annual Report for 1944.

[91] KNA/DC/HRA/2/1, Handing over reports for Garissa District from L. Hale to R.S. Wiener.

short rains were favorable, the long rains failed.[92] The short rains did not sufficiently improve the pasture and water conditions on the British side, so once again, the British pastoralists were forced to cross the frontier. The Moyale District report describes the situation in bleak terms.[93]

The perennial problems of water and grazing scarcity in the NFD created various perceptions about the precarious struggles regarding herd management. The provincial report for the period underscores how the pastoralists viewed the situation: 'Whether a particular year was a satisfactory one depends almost entirely on the amount of rainfall and distribution with resultant grazing.'[94] True to this perception, 1945 was considered a famine, referred to by the local Borana as the drought of hides and skins (*oola qoolaji*). The name indicates the high number of hides and skins in the Moyale markets.[95]

Compared to the previous seven years, the long rains in 1946 were reported as the heaviest. The Moyale District annual report states, 'The Titu lagga [seasonal river] ran in the long rains, and the pools were filled, and the Borana enjoyed good watering and the grazing there [throughout the] year.'[96] This favorable rainfall continued into 1947. The rains came outside the regular season, falling during the dry season and filling the rain pools. This was followed by heavy falls during the long rains. Still, these were concentrated mainly on the Ethiopian side of the frontier, with the British nomads sending much of their dry stock across Ethiopia.[97] The rainfall in 1949 was again low during the long and the short rainy seasons. The rains were scattered, and drought conditions had intensified. The unfavorable water and grazing conditions again resulted in high stock mortalities, forcing herders to trek across Ethiopia.[98] The 1950s experienced rainfall fluctuations like the past—heavy rains interspersed with droughts.

[92] KNA/PC/NFD1/1/8, R. H. Symes Thomson, Northern Frontier District Report for 1945.

[93] HMS 1949, Somaliland Protectorate, 34.

[94] KNA/PC/NFD1/1/8, R. H. Symes Thomson Northern Frontier District Report for 1945.

[95] Interview with Oba Sarite Kura in 1992, Sololo, Kenya.

[96] KNA/PC/NFD1/6/4, Moyale District northern Frontier Annual Report for 1946.

[97] KNA/PC/NFD1/6/4, Moyale District northern Frontier Annual Report for 1947.

[98] KNA/PC/NFD1/1/9, R.G. Turnbull, Annual Report Northern Province 1949.

Heavy Rains and Droughts, 1950–1962

During this period, the main challenge to pastoralism was swarms of locusts that aggravated drought conditions.[99] The rainfalls in 1950 and 1951 were almost a failure, the lowest since 1943. Consequently, some parts of the NFD province experienced near-famine conditions in two different seasons. In contrast, the rains were evenly distributed in others, and in Moyale, the short rainy season 1951 (October–December) started with a dry spell, causing another famine. This was the third time that the population of the NFD received famine relief.[100] The situation differed in the Mandera District, where the long rains of 1951 were reported as 'excessive.'[101] However, this heavy rainfall was associated with a high death toll in the malaria epidemic.[102] As in past years, 1952 was a period of water shortages.[103]

The year 1953 was a repeat of the patchiness of the rains in general, except for the Moyale pastoralists, who considered it one of the better years for livestock production. The administration intended to implement grazing controls, using the Borana as a model group, by moving them south into the lowlands of Golbo, where grazing was plentiful, to preserve the grazing resources along the foothills of the frontier. This was one of the few times when the Uaso, Moyale Borana, and Sakuye shared grazing and water in the region of Bododa.[104] This situation was once again changed in 1954. The March–May–June long rains were moderate, but the following short showers were highly patchy.[105] The year 1955 was generally favorable, as the widespread rains continued into 1956 and 1957. The water supplies in the pans and pools allowed pastoralists to use their allocated grazing lands for extended periods.[106]

[99] KNA/PC/NFD1/1/9, Desmond O'Hagan, Northern Frontier Province Annual Report for 1950.

[100] KNA/DC/MLE/2/2, AGR.11/1943-62, From H. W. Waller the District Commissioner Moyale to Provincial Commissioner Isiolo.

[101] KNA/DC/MDA/1/3, Mandera District-Northern Province Annual Report 1951.

[102] KNA/DC/MDA/1/3, P.M. Gordon, Mandera District-Northern Province: Annual Report 1951.

[103] KNA/PC/NFD1/1/10, R.G. Turnbull, Annual Report Northern Province 1952.

[104] KNA/PC/NFD1/1/10, N.F. Kennaway, Northern Frontier Province Annual Report for 1953.

[105] KNA/DC/MDA/1/3, Mandera district–Northern Province Annual Report 1954.

[106] KNA/DC/MDA/1/3, Mandera district–Northern Province Annual Report 1956.

The situation in 1958 mirrored the previous years when the failures of either the long or the short rains left the pastoralists with inadequate water supplies.[107] In the Moyale District, the long rains were just above average (i.e., more than 350 mm per year). However, the Golbo lowlands were flooded by heavy rains that fell on the Ethiopian side. The floods filled the depressions of Bododa, attracting the region's pastoralist populations from the Isiolo and Moyale districts.[108]

In 1961 the province suffered one of the worst famines in recent history. Many pastoralists died of starvation, and for the first time, the provincial administration received international food aid and supplied canned fish to the nomads. Characteristic of the NFD climate conditions during the previous half-century, the heaviest rainfall fell in 1962 (now understood as the El Niño rains), flooding the lowlands and cutting the province off from the rest of the country. The colonial government supplied relief food packages, dropped from the air with the help of the Southern Rhodesian Air Force.[109] This region's fluctuating wet, dry, drought, and excessive rainfall reflect the history of variabilities and continuities of pastoral resilience.

Variabilities and Continuities of Pastoral Resilience

Regarding the analysis, pastoral resilience might be understood in terms of fluctuations in response to various shocks yet remain persistent for a long time. When dealing with climatic shocks, we have shown that the NFD pastoralists used mobility to maintain resilience by responding to pasture and water's spatial and temporal variations.[110] We have demonstrated how pastoralists responded by crossing the international frontier when necessary. For the herders, they aimed to sustain the resilience of their herds. We can summarize the events from the beginning and end of colonialism on pastoralism by the scheme in Fig. 4.1, which shows how the encapsulation of trans-frontier pastoral communities initially broke down their traditional grazing movement methods. This historical knowledge is critical in exploring ongoing debates on adaptations to disasters and political

[107] KNA/DC/MDA/1/3, Annual report for 1958, Mandera District Northern Province.

[108] KNA/PC/NFD1/1/1/2, P.E. Walters Northern Province Annual Report 1960.

[109] Oba, G. *Climate change adaptions in Africa: An historical ecology* (London, Routledge, 2014).

[110] McPeak, J. G. and Little, P. D., 'Applying the concept of resilience to pastoralist household data,' *Pastoralism: Research, Policy and Practice* 7 (2017): 14.

conflicts. The length of historical time and richness of historical data have allowed us to gain an overview of events and the societies' adaptations to climate variability. The scheme shows how the colonial closing of the frontier broke down the pre-colonial patterns of land use. The pastoralists responded to trans-frontier grazing, risking political pressures. Over nearly five decades, pastoral land use responded to the combined effects of ecological and political shocks.

To illustrate the resilience process, we compare wetness indices as predictors of rainy, dry, and drought years in the NFD (Fig. 4.2). The years with negative indices depict periods of livestock population collapse, while the positive indices indicate herd recovery.[111]

Generally, rainfall patterns in the NFD varied from patchy to totally lacking to heavy rainfall. The heaviest rains occurred 10 to 15 years after

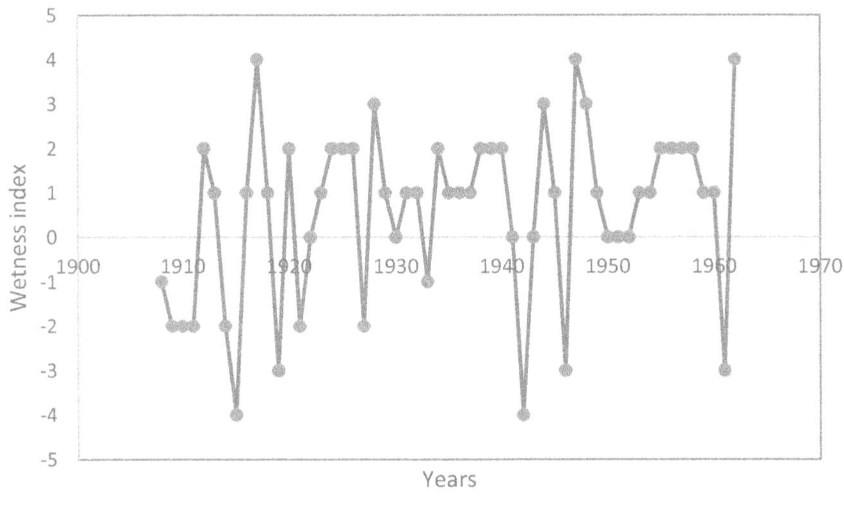

Fig. 4.2 Wetness index of the NFD grazing lands between 1908 and 1962. The indices indicate years of scattered rainfall (0), average rainfall (+1), heavy and very heavy rainfall (+2), extreme wetness (+4), droughts (-1 to -2), and famine (-3 to -4)

[111] Galvin, K. A., Thornton, P. K., Boore, R. B. and Sunderland, J., 'Climate variability and its impacts on east African livestock herders: The Maasai of Ngorongoro conservation area, Tanzania,' *African Journal of Range and Forage Science* 21 (3) (2004): 183–189.

flood waters filled all depressions. Regional and local droughts within NFD reflected different rainfall patterns; on the one hand, some parts of the province experienced near-famine conditions, while in others, the rains were above the long-term average. Rainfall deviations from the long-term mean over 37 years (1935–1962) show droughts occurring 56.8 percent of the time, relative to above-average rainfall that occurred 43.2 percent of the time. Famines were common—major famine occurred four times over six decades. Although these were periods of socio-economic collapses, recoveries were possible, indicating high levels of resilience.[112] There was, however, a change in the pastoralism trajectory after the independence of Kenya in 1963, when the NFD became a scene of secessionist war (Chap. 5).

[112] Oba, G. *Climate change adaptions in Africa.*

PART II

Reorganization and Adaptive Diversities

CHAPTER 5

Collapse and Transformation of Pastoralism: Pathways of Land-use Change, the 1960s–2000s

The decades of the 1960s and 1970s coincided with another collapse of the pastoral economy in Kenya's former Northern Frontier Districts (NFD). In a collapsed rural system, there ought to be a reference to the trigger(s) that stimulated the process. In the mid-twentieth century, the external stimulus that radically changed the trajectories of pastoralism in northeast Africa was warfare (see also Chap. 3). The trigger caused substantial losses in the integrity of the pastoral social-ecological system. The impacts altered pastoralists' lifestyles from nomadism to sedentarization. The changing livelihood patterns dramatically transformed land use from animal husbandry to crop cultivation. Meanwhile, following the initial war trigger, an array of drivers and the overarching environmental variability and human behavior determined the direction of land-use change. The approach reflects adaptive strategies, which depend on the capacity of the socio-ecological system to create new economic domains of the transformed landscapes. The processes marked a new benchmark of social-economic resilience in the agropastoral system.

This chapter describes the political and economic triggers that collapsed pastoralism in Kenya's Northern Frontier District (NFD). The stimuli were secessionist wars promoted by the irredentist policies of Somalia after independence in 1960. The newly independent Somali Republic expressed desires for Somali relatives from across the Ogaden region of Ethiopia, Djibouti, and those inhabiting the part of the NFD in

G. Oba, *Pastoralist Resilience to Environmental Collapse in East Africa since 1500*, https://doi.org/10.1007/978-3-031-48291-5_5

Kenya to rejoin the republic.[1] In Kenya, the districts of Mandera, Wajir, and Garissa (inhabited mainly by Somali clans) and the non-Somali sections of Isiolo, Moyale, and Marsabit were regions the republic claimed as part of its territory.[2] Shortly after Kenya gained independence in 1963, the pro-secessionist parties in the NFD joined the guerrillas of the Northern Frontier District Liberation Army (NFDLA)[3] to fight against security forces and ethnic groups that were opposed to secessionism.[4] In October 1963, the bandits—locally called *shifta*—were on the offensive, attacking government outposts and civilians.[5] The Kenyan security forces' predatory raids and counter-insurgency initiatives caused massive loss of lives and disrupted traditional land-use patterns.[6] The pastoralists were placed in security settlements, a period which the local pastoralists referred to as a 'stop' that described the collapse of pastoralism.[7] The camel economy was wholly shattered—hence, the economies dominated by camels entirely collapsed.[8] The small stock was virtually wiped out, while for cattle groups, the majority were impoverished.[9] We will use the Obbu Borana—the group most adversely affected by the war—to investigate the collapse of pastoralism that transitioned into land-use change. Two processes that led to land-use change include—the choice of settled agriculture and the enclosure of communal grazing lands to preserve grass pastures for private

[1] Drysdale, J., *The Somali dispute* (New York: Fredrick A. Praeger, Inc., 1964). Ichani, X. F., Matheka, R. M. and Wario, H. A., 'Debating on success of the Kenya counter insurgency on the shifta war: Between double success and little consolation,' *International Affairs and Global strategy*, 71 (2019). 10.71.76LAGS.

[2] Robinson, C., 'Revisiting the rise and fall of the Somali armed force, 1960–2012,' *Defense and Security Analysis* 32 (3) (2016): 257–252.

[3] Branch, D., 'Violence, decolonization and the cold war in Kenya's North-Eastern Province, 1963–1978' *Journal of Eastern African Studies* 8 (4) (2014): 642–657.

[4] Ibid.

[5] Ichani, Xavier Francisi. *Kenya defense forces and militarization of internal disputes, 1963–2013.* Diss. Egerton University, 2019.

[6] Whittaker, H., 'Pursuing pastoralists: The stigma of shifta during the "shifta war" in Kenya, 1963–68,' ERAS Edition 10. www.arts.brash.edu.au/publication/eras.

[7] Khalif, Zeinabu Kabale, and Oba, G., '"Gaafa dhaabaa-the period of stop": Narrating impacts of shifta insurgency on pastoral economy in northern Kenya, c. 1963 to 2007,' *Pastoralism: Research, Policy and Practice* 3.1 (2013): 1–20.

[8] Hjort, A., *Savanna town: Rural ties and urban opportunities in northern Kenya* (Stockholm, Studies in social anthropology, 1979).

[9] Hogg, R., *Re-stocking pastoralists in Kenya: A strategy for relief and rehabilitation*, ODI, PDN 19C, 1985.

use.[10] Historically, the Borana pastoralists were not farmers. They shared a grain-labor-client relationship with farming communities in southern Ethiopia. We examine historical linkages in farming technology south of the Ethiopian–northern Kenyan frontier. Additionally, we will consider a variety of drivers that directly and indirectly triggered shifts to agropastoralism.[11] We will show how this relationship with Ethiopian farmers spurred the community's adoption of farming.

Structurally, land use shifted from communal grazing to private.[12] Economically, farming and range enclosures supplied income and diversified livelihoods.[13] The purpose is to ensure food security for designated groups or individuals. Indeed, there is a hypothesis that self-grown crops and preserved pastures would supply better livelihood alternatives than declining herds.[14] The challenge for new farmers was the integration of crop cultivation and stock management under conditions of environmental variability. According to one herder, the decision by the Obbu Borana households to take part in these activities was to have 'one hand on the plow and the other holding the cow's tail.' The choice depended on climate conditions—abandoning farming and holding onto cattle during droughts while shifting resources and labor to agriculture when the rains returned. We will evaluate outcomes and the extent to which the collapse of pastoralism and the emergence of farming and range enclosures contributed to agropastoralism's social and economic resilience. We would investigate the specific constraints experienced during socio-economic changes and show how adopting farming practices and technologies results in significant changes in the livelihoods of the Obbu Borana households.

[10] Renom, J. G., Mwamidi, D. M. and Domi'nguez, P., 'Holistic ethnographies of East Africa customary pastoral commons needed?' *Current Opinion in Environmental Sustainability* 43 (2020): 83–90.

[11] O'Riordan T. and Lenton, T., *Addressing tipping points for a precarious future*, Published for the British Academy by Oxford University Press, 2013.

[12] Ca'ceres, D. M. and Gras, C., 'A tipping point for agricultural expansion? Technological changes and capital accumulation in Argentina's rural sector,' *Journal of Agrarian Change* 20 (2020): 79–93.

[13] Burian, A., Karaya, R., Wernersson, J. E. V., Egberth, M., Lokorwa, B. and Nyberg, G., 'A community-based evaluation of population growth and agro-pastoralist resilience in Sub-Saharan drylands,' *Environmental Science and Policy* 92 (2019): 323–330.

[14] Fratkin, E., 'East African pastoralism in transition: Maasai, Boran, and Rendille cases,' *African Studies Review* 44 (3) (2001): 1–25.

To understand transitional pathways of land-use change, we pose the following questions: (a) How did the war-triggered collapse of pastoralism and the transition of agropastoralism frame land-use change? (b) To what extent did the history of grain-labor-client relationships with farmers in southern Ethiopia influence the Obbu Borana to adopt crop cultivation? (c) What was the performance of farm production at household and locational levels? (d) How did range enclosures supply income opportunities? (e) How may we explain the resilience of agropastoralism under changing land-use conditions? We address each of these questions in turn.

How Would the Collapse of Pastoralism and Transition of Agropastoral Land Use Be Framed?

The collapse placed acute stress on the livelihoods of former nomadic pastoralists. The process triggered structural changes in social-ecological systems, stimulating innovative adaptative strategies. The society transitioned from the socially homogenous pastoral subsistence to a complex and diversified agropastoral system.[15] The quantitative thresholds crossed are in terms of transitions in land use bringing about long-lasting changes in social-ecological systems.[16] Conversely, land-use transitions triggered the fragmentation of the communal grazing lands into semi-private farms and range enclosures.[17] These are described in terms of the adaptive cycle—applied to transitional land use (Fig. 5.1).

Assessing the impacts of war on land-use transformation presents complex scenarios and pathways.[18] In the 1950s (pre-independence years), the conditions (pathway A) represent a resilient pastoral land-use system. By the early 1960s, regional wars had collapsed land use (pathway B). The war displaced pastoralists from more than 80 percent of the grazing lands

[15] Fratkin, E., 'East African pastoralism in transition: Maasai, Boran, and Rendille cases,' African Studies Review 44 (3) (2001): 1–25.

[16] Cumming, g. S. and Peterson, G. D., 'Unifying research on socio-ecological resilience and collapse.'

[17] Galvin, K., 'Transitions: pastoralists living with change,' *Annual Review of anthropology* 38 (2009): 185–198.

[18] Serneels, S. and Lambin, E. F., 'Proximate causes of land use changes in Narok District, Kenya: A spatial statistical model,' *Agricultural Ecosystems and Environment* 85 (2001): 65–81.

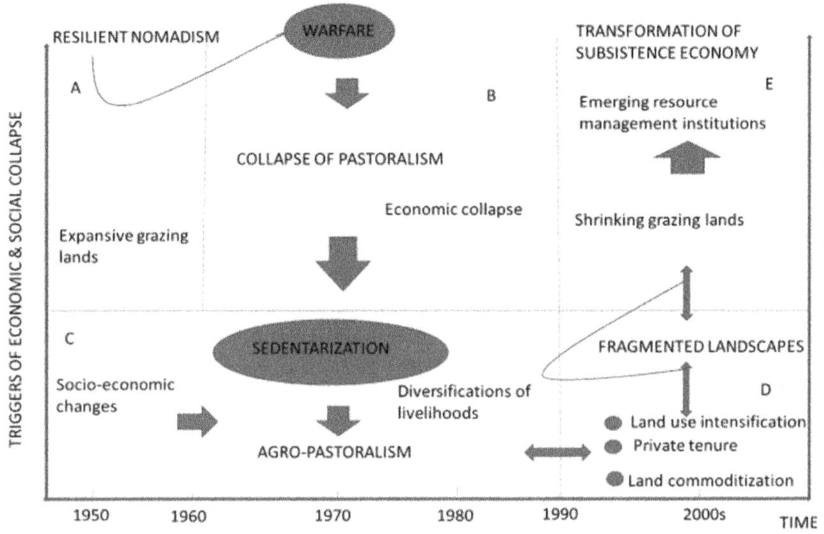

Fig. 5.1 Pathways (A–E) showing the collapse of pastoralism, socio-economic transformation, and land-use changes. Changes depicted by pathway E are discussed in Chap. 6

in northern Kenya, destroying the pastoral economy.[19] The war forced the former pastoralists into settled agriculture (pathway C). The far-reaching changes occurred in land tenure (pathway D). The privatization of land tenure was motivated by future monitorization.[20] Consequently, the tipping points of land-use change were not only a threat to the communal resources but also offered potential moments by redirecting future land-use arrangements.[21] The consequences of shrinking grazing lands to land-use adaptive strategies (pathway E) will be discussed in Chap. 6. The processes combined motivated agropastoralism.[22]

[19] Ibid.

[20] Tynon, Joanne F., Sundstrom, S. and Western, D., 'Rangeland Privatization and the Maasai Experience: Social Capital and the Implications for Traditional Resource Management in Southern Kenya' (2012).

[21] O'Riordan, T. and Lenton, T., *Addressing tipping points for a precarious future* (Oxford: The British Academy by Oxford University Press, 2013).

[22] Little P.D., Smith, K., Cellarius, B. A., Coppock, D. L. and Barrett, B., 'Avoiding disasters: Diversification and risk management among East African herders,' *Development and Change* 32 (2001): 401–433.

Agropastoralism comprises three components: the natural ecological, socio-economic, and animal/crop production systems.[23] Regarding management, biological, environmental, and socio-economic systems logically represent the same landscape—one having natural resources and the other having management features. Managing natural ecological and socio-economic systems has evolved complex interactions by linking crop cultivation with animal husbandry.[24] Accordingly, when a farmer states, 'one hand on the plow, the other holds a cow's tail,' it is a metaphoric description of the multiple goals of managing crop-animal husbandry production.

In semi-arid lands, where crop and range production are negatively influenced by climate variability[25]—rains that are sufficient for growing grass might be insufficient for growing crops; hence, the local terminologies—'grass' and 'crop' rains.[26] The implication is that there are clear cycles of agricultural production interrupted by brief attention to herd management. Still, droughts induce collapses in herd demography[27] as well as crop failures. So, adaptations are *ad hoc* dependent on the severity of droughts.[28] Farming fluctuates between gains and losses.[29] Still, when conditions are favorable, agriculture supplies a vital food source. Households might sell surplus grains or share them with friends and

[23] Dong, S., Liu, S. and Wen, L., 'Vulnerability and resilience of human-natural systems of pastoralism worldwide' in Dong, Shikui, Lu Wen, Shiliang Liu, Xiangfeng Zhang, James P. Lassoie, Shaoliang Yi, Xiaoyan Li, Jinpeng Li, and Yuanyuan Li. 'Vulnerability of worldwide pastoralism to global changes and interdisciplinary strategies for sustainable pastoralism.' *Ecology and society* 16, no. 2 (2011).

[24] Eneyew, A., 'Is settling pastoralists a viable livelihood strategy? Implication for policy dialogue,' *Scholarly Journal of Agricultural Science* 2 (5) (2012): 94–102.

[25] Shukla, S., McNally, A., Husak, G. and Funk, C., 'A seasonal agricultural drought forecast system for food-insecure regions of East Africa,' *Hydrology and Earth System Sciences* 18.10 (2014): 3907–3921.

[26] Mace, R., Anderson, D. M., Bierschenk, T., Cronk, L., Köhler-Rollefson, I., Lancaster, W., Lancaster, F., Little, P. D., Ann Morris, E. and Rossignol, J., 'Transitions between cultivation and pastoralism in sub-Saharan Africa [and comments and reply],' *Current Anthropology* 34 (4) (1993): 363–382.

[27] O'Connor, T. G. and Kiker, G. A., 'Collapse of Mapungubwe society: Vulnerability of pastoralism to increasing aridity,' *Climate Change* 66 (2004): 49–66.

[28] Speranza, Chinwe Ifejika. 'Drought coping and adaptation strategies: Understanding adaptations to climate change in agro-pastoral livestock production in Makueni district, Kenya' *The European Journal of Development Research* 22, no. 5 (2010): 623–642.

[29] Watete, P.W., Makau, W. K., Njoka, J. T., Opiyo, L. A. and Mureithi, S. M., 'Are there options outside livestock economy? Diversification among households of northern Kenya,' *Pastoralism: Research, Policy and Practice* (2016) 6: 3. 10.1186/s13570-016-00504.

relatives, while during periods of drought, society relies on relief food (see Chap. 9). Additionally, fencing the commons created private pastures.[30] The consequences are examined regarding the emerging monitorization of the agropastoral system. We begin the analysis with historical events that prompted the Obbu Borana pastoralists to adopt farming.

To What Extent Did the History of Grain-labor-client Relationships with Farmers in Southern Ethiopia Influence the Borana to Adopt Crop Cultivation?

Before the *shifta* war in the 1960s and 1970s, the Obbu Borana[31] considered breaking the earth a cultural taboo.[32] Nonetheless, grain was important in household diets, especially during drought years. During this earlier period, the Obbu residents formed grain-labor-client relationships with Ethiopian farmers from across the frontier. We interviewed elders who knew of that time and those activities. Farming on the Ethiopian side of the border began in the last decade of the 1890s after the conquest of the region by rifle-bearing farmers— a vanguard of Menelik's armies.[33] In the southern frontier, the settlers grabbed vast lands called *Qalada*. The Borana contributed labor for weeding and harvesting and received specific amounts of grain for their services given—the relationship evolved into a 'grain-labor clientship.'[34] Local narratives describe such relationships shared with prominent Ethiopian farmers.

[30] Debsu, D. N., 'Customary always in Ethiopia: Institutions of dispute resolution among the Guji Oromo agropastoralists,' *Journal of Oromo Studies Association* 17 (2) (2010): 101–127.

[31] By 1962, only two shopkeepers were known to cultivate plots of land. The shopkeepers were Guyo Huqana Ch'aari, and Dido Guyo.

[32] Baxter, Paul TW. 'Ethiopia's unacknowledged problem: the Oromo,' *African Affairs* 77, no. 308 (1978): 283–296.

[33] The sections of the Borana in the sub-humid zone, might have begun farming during earlier centuries but their farms were destroyed by the orders of the gada (Oba, G. 1996, 'Shifting identities along resource borders: Becoming and continuing to be Borana Oromo,' in Baxter, P.T.W., Hultin J. and Triulizi A. (EDS.) *Being and becoming Oromo: Historical and anthropological enquiries.* Uppsala, Nordiska Africainstitutet.

[34] Oba, G., 'Changing Property Rights among Settling Pastoralists: An Adaptive Strategy to Declining Pastoral Resources,' In PTW Baxter (ed.) *Property, Poverty, and People: Changing Rights in Property and Problems of Pastoral Development* (1994): 38–44.

Among the well-known settler farmers (early twentieth century) around Tuka was Rafeela *Abba Qiita*. He was a wealthy landowner and politically well-connected. Around Gombisa, also in southern Ethiopia, where the likes of Baaru Gaadisa (*Abba Maagal*[35]), because of the considerable extent of his farms, created local myths about wealth,[36] a common satirical joke among friends was '*aatini Baaru uufi seeta?*'—'You think you are Baaru.' He was the epitome of large farm owners who required hundreds of trained pairs of oxen for plowing and thousands of people to do the weeding and harvesting.[37] According to one of our informants, after harvests, the Ethiopian farmers threshed grains (maize) and stored them in underground pit silos called *bankara*.[38] Huqa Morowa shared his experience:

> The *Sidama* farmers stored grains in underground silos to save themselves from recurrent famine. They dug chambers several meters deep and "fired" the inside to make the soil firm and sterile. They then applied wet cattle dung to plaster the walls of the silos… smoked until the plaster dried. After thrashing from the cobs, the harvested maize grains were measured in the silo chambers. [When done], the external opening is wholly sealed [with rock slabs and plastered with clay mixed with cow dung] so that soil does not contact the grains. [After long storage], the hot temperatures in the pit silos killed seed-eating weevils and grain fungus. The silos were opened only during periods of severe food shortages. These were the times when the Obbu Borana visited their hosts…. The maize …given on credit, the terms were a bag of maize (i.e., 100 kg) bartered for a bull.[39]

[35] Taking the names of their horses—the father of grey bull horse.

[36] Interviews Rev. Shedo Oba and the late Huqa Morowa by Gufu Oba, in August 1996.

[37] There is women maize song that goes as follows: *Ka Baaru Magalla, kadaatu badassa-Baruni galaana guumi walqitessa. Waalaten bineensa entu shoka nu diesa*,—directly translated—the brown grains of Baru feeds beggars. Baru is generous like a river, and he equalizes all, while his wife Walate is a beast, who can provide us with an ax! The reason was that she took back the grain given to the beggars. She only allowed those who worked on the farms to collect poor maize cobs left in the field. Interview with Godje Jillo on 12 July in 1996.

[38] McCann, James C. *People of the plow: An agricultural history of Ethiopia, 1800–1990.* Univ of Wisconsin Press, 1995. James McCann suggests that the system of burying grains in pits originates from the north and serves as effective means of storing grains at storage temperatures that would be unfavorable for weevils to affect the grain, while the pits were also used for strategic reasons for they could easily be hidden and plowed and planted over to avoid looting by tax collectors (p. 68).

[39] Interview with Huqa Morowa by Gufu Oba, in August 1996.

During dry seasons when the pit silos were opened, the information was broadcast by word of mouth throughout the region. Pastoralists from as far as the plains of northern Kenya would arrive to buy grains and benefit from credit facilities. Individual Borana families established a clientele relationship with the settler farmers to access long-term grain credits after producing a guarantor (*waasi*)—usually a prominent clan elder. The terms of credit were specific measures in *ferasila*[40] paid with interest on the unit value. For example, 100 kg (5 *ferasila*) of grain would be repaid after a year, with double the value of its weight (i.e., 100 × 2 or 200 kg of grain) at the current market price on the principal.[41] Accordingly, for the Ethiopian farmers, grains were more than just food for immediate consumption—they were a form of productive capital that served as an investment with good returns if planted. As an extension of the feudal government, the Ethiopian settlers could enforce payment in case of defaulting—confiscating the livestock of the clan member who had provided the collateral.[42] For the Obbu households, the deal had certain benefits—they did not have to sell their cattle assets when animals were in poor condition. The herders benefited by paying their debts when the stock gained weight and fetched better prices at the market.

By the 1970s, the grain-labor-client system had ended—after a century. At that time, attacks by the Somali Abbo Liberation Front (SALF) operating across the Ethiopian frontier targeted the Ethiopian settler farmers, forcing them to abandon their farms.[43] Although the war ended the historical grain-labor-client relationship, the experience influenced the Obbu Borana to adopt farming practices and technologies.[44] In describing pathways related to socio-economic changes (pathway C), we investigated the performance of crop cultivation and the constraints experienced.

[40] *Farasile* or *farasila* is an Ethiopian measure around 17–20 kg; it varies from region to another.

[41] Personal discussions with the late Bidhu Bankare, July 1990.

[42] Interview with Rev. Shedo Oba Kuricha by the author in 1994, in Sololo.

[43] Interview with Paul Guyo Bonaya in Anona on 19 July 2010 and Huqa Morowa Kaar Bururi on 2nd August 1996.

[44] Technologies such as grain storage (*gootara*), pairs of plowing oxen (*qiindi*), plow (*meresha*), yoke (*qampara*), plow stick (*hiirfi*), plow board (*batate*), and the whip (*qaache*) are adoptions from the Ethiopian farmers.

What Was the Performance of Farm Production at Household and Locational Levels?

Crop cultivation is conducted along a narrow strip bordering the Ethiopian southern escarpment (see Fig. 2.3). Crop cultivation began in 1962[45] and continued after the drought of 1970–1973.[46] The process starts with clearing, burning, and fencing land. Farmers used the most straightforward tools, such as the hand hoe, during the first phase to plant crops. This cultivation phase is the least productive, as weeds tend to subdue the growth of crops. Later, farmers used trained oxen and hired labor to improve crop cultivation, gaining better control of weeds and increasing crop yields.

According to the households we interviewed in 1987 ($N = 164$), 92.8 percent reported having farms, compared to 7 percent that had none. Only a tiny proportion of the households surveyed claimed to have bought, inherited, or used personal initiatives to establish farm ownership (Table 5.1). There were different rules governing access to land for cultivation compared to grazing. The farms were not adjudicated, so the rights

Table 5.1 Methods of establishing ownership rights to farmlands ($N = 164$)

Methods of establishing land ownership	*% Response*
Chiefs' committees	66.7
Purchase	3.9
Inheritance	11.1
Own initiatives	9.8
Use other people's land	3.2
Borana elders	2.6
Agricultural officers	1.3
The administration	1.3

[45] The pioneers were Guyo Huqana Chaari and Dido Guyo, both, traders in Sololo-Ramata. Using the labor from the Ethiopian side of the border to train plow oxen and manage the farms. The author as a youth remembers hearing negative comments by the herders visiting the town, when the hired laborers, among them, a man called Lubo, used whips to keep the oxen pull the wooden plow. 'See what the enemies of cattle do to them!' was a common comment.

[46] Oba, G., 'Changing Property Rights among Settling Pastoralists: An Adaptive Strategy to Declining Pastoral Resources,' in PTW Baxter (ed.) *Property, Poverty, and People: Changing Rights in Property and Problems of Pastoral Development* (1994): 38–44.

Table 5.2 A number of years involved in farming across locations in the Obbu region

Locations	*% Response Years of farming*					
	0–5	*6–10*	*11–15*	*16–20*	*21–25*	*>26*
Anona	0	10	20	10	30	30
Mukutano	33.3	0	16.7	33.3	16.7	0
Ramata	52.6	15.8	5.3	15.8	0	10.5
Madho Adhi	10	10	20	20	30	10

must be legally fixed. The Borana were aware that the government's agricultural policy encouraged private ownership of farmlands. According to the interviewed households, farming provided diversified livelihoods (30.8%), helped them to avoid buying food (7.6%), and reduced direct reliance on cattle (38.5%), thereby providing opportunities for the herds to grow (23.1%).

Farming expanded in the 1980s and the 1990s—particularly after the 1983/84 and 1991/92 droughts. Among the households, the years they were involved in farming varied from less than 5 to more than 26 years (Table 5.2). On average, 64.8 percent of the farms were less than 5 acres.

The relationship between farms and family sizes was positive (Pearson's $r = 0.185$, $p = 0.027$), implying that more prominent and wealthier families tended to have larger farms. Smaller and poorer families, accordingly, had smaller farms.[47] This was true at the time of the survey, while a decade later, all suitable farming lands had been occupied. Whether farming succeeded or failed, the goals of food self-sufficiency depended on environmental, economic, and technical factors that constrained farming.

Constraints on Farming

The critical constraints on farming included climate variability. Rainfall in the region is bimodal, with the long rains expected in March to May (*ganna*) and the short rains (*hagaaya*) in late October through December.

[47] Tache, Boku, and Gufu Oba. "Is poverty driving Borana herders in southern Ethiopia to crop cultivation?" *Human ecology* 38, no. 5 (2010): 639–649.

Planting begins in mid-March and continues after the first rain showers in April. The maize crop ripens in June, with harvests in July. During the short rains, crops are planted at the end of October and early November and, if successful, are harvested in late December to early January. Better harvests were expected during the long rains, while yields were likely to fail during the short rains. Accordingly, unless the short showers were successful, farming remained subordinate to livestock management at that time of the year. The informants described rainfall as unpredictable, insufficient in amount, and uneven in the distribution in space and time. When rain was perceived inadequate for growing crops, the planted fields were temporarily abandoned, with the emphasis shifting to cattle management. Since rainfall is unreliable, crop farming remains opportunistic. The goal of every farmer was to capture the moisture of the first rainfall by early planting. So, although traction power improved yields, the climate had an overarching effect on crop production. For Obbu households, this study estimated that a successful harvest occurred once every four years. The sensitivity of crop varieties to climate variability and low soil fertility would explain poor farming performance. Additional constraints on successful farming were the lack of certified seeds, inferior farming methods, problems with weeds, the lack of traction animals, inappropriate farming tools, and a lack of fertilizers (Fig. 5.2).

Additionally, labor shortage was a constraint on successful farming. About 68 percent of the households claimed they used communal labor to plant and weed their neighbors' crops. Other families (32.2%) did not benefit from cooperative work. Following the droughts of 1983/84, these systems of neighborhood aid were replaced by hired labor. In the 1980s, individuals were paid KSh 50–70, working seven hours daily.[48] By the 1990s–2000s, the labor price for seven hours had increased to KSh 500 per person per day. During favorable crop-growing years, high labor prices motivated poor farmers to sell their labor as an option rather than cultivating their farms.[49] The commercialization of farm labor meant that only salaried individuals, business owners, and wealthy households could afford to hire work (5.1%). In comparison, the majority (94.9%) had no financial capacity to pay for hired labor. Like labor, access to trained pairs of oxen

[48] At the time this work was written in 2021, the charges for hiring labor for the same number of hours has increased to KSh 500 per day per person.

[49] Shipton, P., 'African famines and food security: Anthropological perspectives,' *Annual Review of Anthropology* 19 (1990): 353–394.

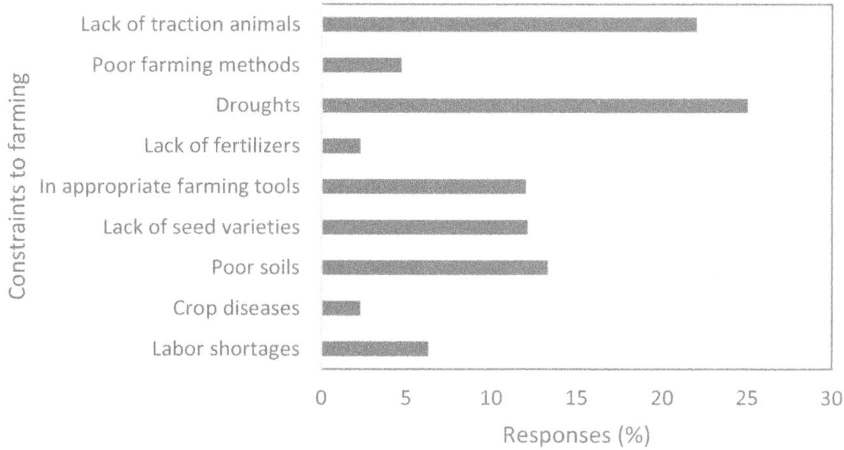

Fig. 5.2 Constraints on agricultural production ($N = 164$)

for plowing was crucial for successful crop cultivation. During the post-drought period of 1992, 35.6 percent of the households used traction animals, and 45 percent employed a hand hoe, compared to 19.4 percent which used both techniques. Given these limitations, farming performances have been unsatisfactory.

Performance of Farm Production

Crop production was more successful during the long rains (91.98%) than the short rains (18.9%). We can refer to two events. During 1990, the Obbu region enjoyed bumper crop harvests after heavy rains, while during the 1996 long rains, maize crops did not show bumper yields. In household surveys (1996) across nine sub-locations, on average, 76 percent of households harvested less than ten bags (of 50 kg) of maize (Fig. 5.3).

The households reported high variability in crop yields, depending on the technology used. For those households that hired trained pairs of oxen, the costs averaged KSh 575 per day, totaling KSh 1688 per planting season per household. Grain harvests by the Obbu households for 2004–2008 varied considerably. Farms plowed using trained pairs of oxen yielded from seven to eighty bags (each of 50 kg) per season. Grain production varied from three to sixty bags per season for those using hand hoes alone. Apart from the Anona site that received flood waters from the

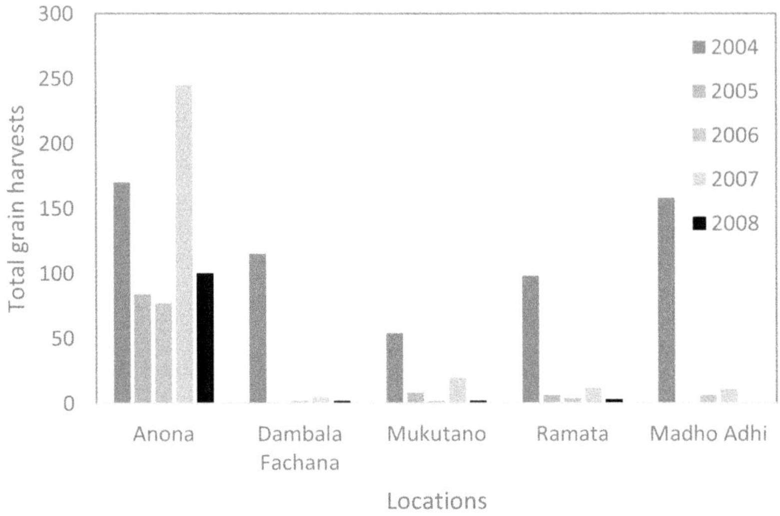

Fig. 5.3 Grain production (numbers of 50 kg bags) dynamics from 2004 to 2008 across five locations

Ethiopian escarpment, the other locations produced low grain yields, apart from 2004. The harvested grains were saved for home consumption, sold, or shared with relatives and friends (see Chap. 9).

Grain Sales Versus Consumption

Of the 164 households interviewed in 1987, on average, 69.5 percent relied on grains for consumption alone, while 30.5 percent depended on grains for consumption and sales. No families cultivated crops solely for sale. Of those who used grains for consumption and sales, more significant amounts were distributed for home consumption and less for sales. Slightly more than half the households (52.9%) shared grains with neighboring households, and most farmers (86.9%) did not save seeds for next season's planting. Self-sufficiency in grains for home consumption (62.3%) lasted four to six months during successful harvests. During dry years, households were food self-sufficient for less than three months. In our 2010 study in southern Ethiopia, crop production supplied 26.3 metric tons per household per year. The yields amounted to 94.6 kg person^{-1} per year or 7.9 kg person^{-1} month^{-1}. This amount provided only 26 percent of the annual grain requirements per person.[50] The surveyed households relied on market sources for added grain supplies. The most significant number of households (96.7%) reported increases in grain prices at the local markets.[51] The price increases in 1987 varied from 233 percent to 500 percent compared to pre-1983 drought prices.

Individual Household Cases

Intra-household differences might be hidden if the data was analyzed across the region.[52] We had the opportunity to examine if there were farmers whose successes might point to their adaptive farming methods. Additionally, the experiences might be general enough to be applied to farming households in the region. From the information gathered, we hoped to interpret if experiences by individual households were informed by gender and management strategies. Understanding if personal farm productivity reflected climate variability would also be possible.[53]

Undoubtedly, individual experiences were uniquely related to household wealth, including financial capital. For example, farmers with trained pairs of oxen were more successful than those using hand hoes. Additionally, access to labor reflected individual successes. Also noteworthy was the fact that individual farmers might have introduced innovative methods.[54] Accordingly, by focusing on individual cases, we hoped to disaggregate different adaptive strategies that might reflect past experiences, economic capacities, and gender.[55] We used four households (case studies 1 to 4) to illustrate other occasions over seven years (1985–1992)—the household of Lokho Halakhe (Cs1), originally resident in southern Ethiopia, crossed into Kenya to escape attacks by *shifta* bandits during the 1970s. She had cattle and small stock and cultivated a plot on which she grew maize. She

[50] Tache, B. and Oba, G., 'Is poverty driving Borana herders in southern Ethiopia to crop cultivation?'

[51] Minority of the households reported no significant changes in grain prices (3.3%).

[52] Tittonell, P., Maruiki, A., Shepherd, K. D., Mugendi, D., Ksizzi, K. C., Okeyo, J., Verchot L., Coe, R. and Vanlauwe, B., 'The diversity of rural livelihoods and their influence on soil fertility in Agricultural systems of East Africa: A topology of smallholder farms,' *Agricultural Systems* 103 (2010): 83–97.

[53] Slegers, M.F.W., '"If only would it rain": Farmers' perceptions of rainfall and drought in se-arid central Tanzania,' *Journal of Arid Environments* 72 (2008): 2106–2123.

[54] Pannell, D. J., 'Social and economic challenges in the development of complex farming systems,' *Agroforestry Systems* 45 (1999): 393–409.

[55] Verma R. *Gender, land, and livelihoods in East Africa: Through farmer's eyes*, International Development Research Centre, 2001.

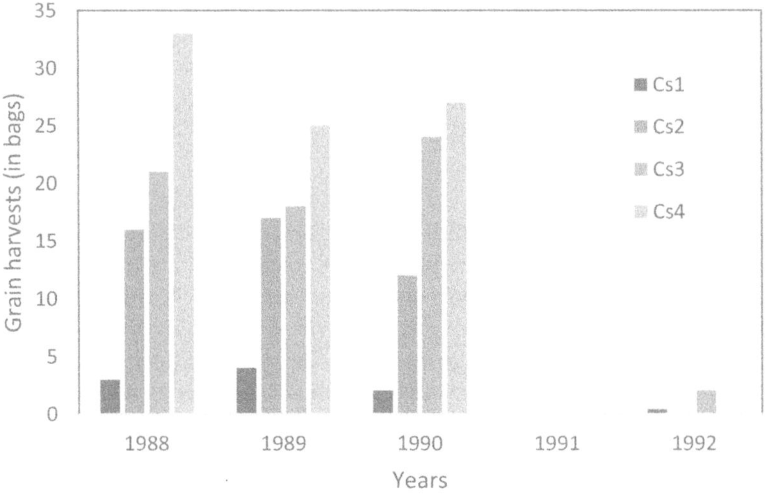

Fig. 5.4 Grain yields reported by four case households for 1988–1992

used hand hoes as farming tools. During the selected five years (1988–1992), her grain harvests were least and varied (Cs1, Fig. 5.4).

The experience of this household can be scrutinized from three perspectives. First, she needed more labor. Second, a lack of economic ability could have influenced her farming. Third, she needed more farming tools. However, yields reported in other years during the period require an explanation. Ignoring the risk of crop failures in 1992, she planted her plot by taking advantage of the early rain showers, which supplied sufficient soil moisture to grow crops despite a few rainy days and the lowest harvest. Here, we might have a case for how individual farmers took risks to exploit the ecological variation of the landscape.[56]

The second case (Cs2) was the household of H'aarero Ulufa. Her family also lived on the Ethiopian side of the frontier before banditry forced them to cross into Kenya. Her family used traction animals to plow the land and planted crops using broadcast techniques. Her harvests during the first three years of the review period were less variable (Cs2, Fig. 5.4) compared to those of others. She obtained stable crops between 1988 and 1990 but none between 1991 and 1992, a drought period.

[56] Pannell, D. J., 'Social and economic challenges in the development.'

We found strong relationships between farming and animal husbandry in our subsequent two household cases (both headed by men). The case households were influenced by their historical experiences.[57] The family of Diba Liban (Cs3) was affected by the tragic war of the 1970s that collapsed the pastoral economy. From 1988 through 1990, the family had moderately successful crop harvests, but these failed during the drought year 1991. Compared to other case families, they had limited yields in 1992 (Cs3, Fig. 5.4). The crop harvest in 1992 might be a reflection of moisture management at landscape scale.

Our next case is the household of Abdub Waqo (Cs4). Before the 1970s, the family lived in Dukale on the Ethiopian border, where ecological conditions were favorable for crop cultivation. The family was forced out of Dukale due to insecurity in 1973. After the war, the family lived on the Kenyan side of the border while he continued cultivating his Ethiopian farm. He plowed his land before the rains and planted crops in rows (*shalshalo*) by the broadcast method while the plants were still seedlings. This technique helped spread the seedlings and increased water infiltration into the root zone. Weeds uprooted by the plow were removed by hand. His harvests from 1988 through 1990 were superior to other households (Cs4, Fig. 5.4). However, during the drought of 1991/92, Abdub Waqo lost his traction animals and was forced to use hand hoes for cultivation, followed by a loss of crop production.

In summing up the cases, the two female-headed households (Cs1 and Cs2) reflect different experiences. Whereas Cs1 worked her farm using hand hoes, Cs2 trained draft animals for plowing, explaining her excellent harvests between 1988 and 1990. In the two male-headed households, their combined long-term experiences in farming and innovative technologies define their more successful crop production. Yet, when climate conditions deteriorated, their crops also failed. The most significant impact of agriculture on agroecosystems is related to how range enclosures influence land use and provide new economic opportunities. We were interested in understanding how range enclosures influence (a) income opportunities and (b) the risk of conflicts. Working on pathway D (Fig. 5.1), we will

[57] Kristjanson, P., Neufeldt, h., Gassner, A., Mango, J., Kyazze, F. B., Desta, S., Sayula, G., Thiede, B., Förch, W., Thornton, P. K. and Coe, R., 'Are food insecure smallholder households making changes in their farming practices? Evidence from East Africa,' *Food Science* 4 (2012): 381–397.

explain how changes in land tenure and the introduction of range enclosures provided additional economic opportunities to the farmers.

How May We Explain Range Enclosures Providing Income Opportunities for Farmers?

Socio-economic factors affecting households and climate variability might have contributed to the development of pasture reserves. Various investments in 'building' ownership rights might be necessary for the *de facto* property rights holder. In our analysis, we use historical approaches to gauge processes of resource use transformation and understand the expansion of range enclosures.[58] The chronological development of enclosures reflects the changes in land-use history (Fig. 5.5).

Before the war in the 1970s, the Borana conserved pastures around settlements as a communal resource referred to as 'calf-grazing reserves.' The establishment of private range enclosures was a negation of this

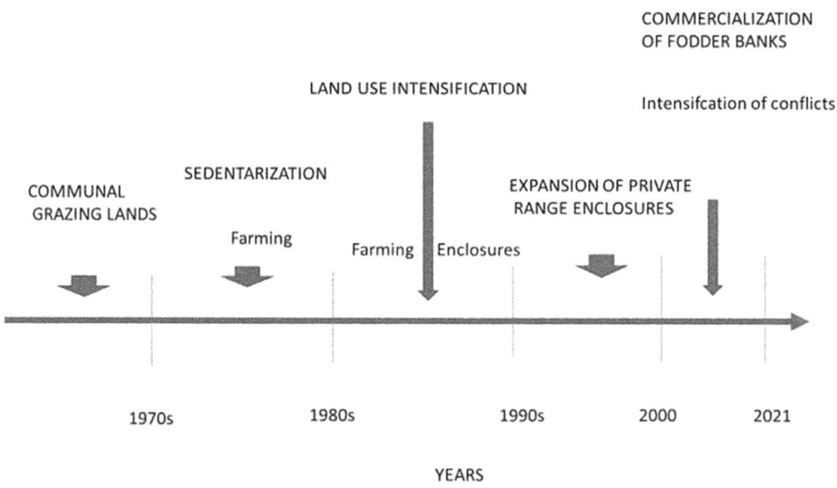

Fig. 5.5 Chronological development of private range enclosures in the Obbu region

[58] Basset, J. J., 'The political ecology of peasant-herder conflicts in the northern ivory coast,' *Annals of the Association of American Geographers* 78 (3) (1988): 453–472.

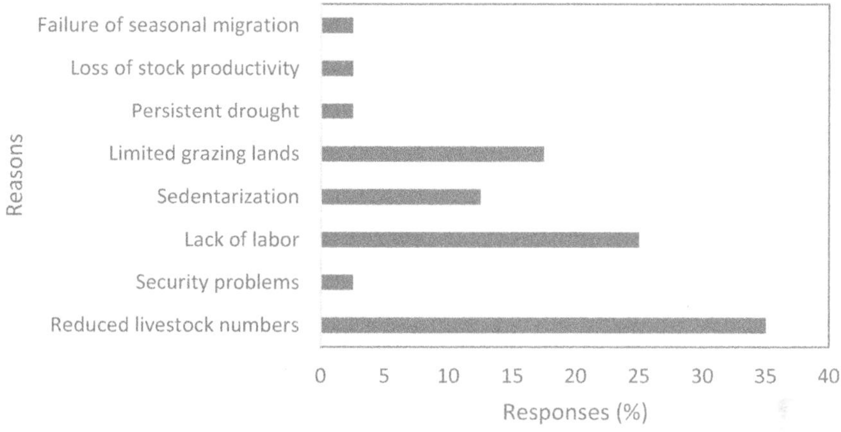

Fig. 5.6 Reasons for establishing private range enclosures

traditional practice. According to our 2008 surveys, there were reasons for establishing private range enclosures (Fig. 5.6). Those who used range enclosures practiced seasonal grazing, albeit on a micro-scale. Depending on the size of the herds, grazing lasted for one to two months before the cattle were moved back to the open grazing areas.

Because grazing was intermittent, the range enclosures accumulated biomass during the growing season.[59] The preserved grass met various needs: it supplied fodder for calves, old and drought-weakened cows (20.1%), and traction animals (20.6%). Other uses were for thatching huts (13.2%), while for poor households, the standing hay was rented by wealthy stock owners (2.6%) at a fee for grazing their livestock.

During the interviews, many informants (93.8%) agreed that private range enclosures were associated with livestock grazing scarcity. The livestock traveled 6–18 km from the homestead daily to search for grazing. The solution by most pastoralists (74.2%) was to establish private range enclosures, while a quarter of them (25.8%) had not done so. The preserved grass pastures were used in three main ways. The first use was to provide grazing for their stock during that critical period of the year when the communal grazing lands were overgrazed. The second use was to rent

[59] Oba, G., Vetaas, O. R. and Stenseth N. C., 'Relationships between biomass and plant species richness in arid zone grazing lands,' *Journal of Applied Ecology* 38 (2001): 836–845.

out the enclosed pastures (done by only 2.6% of the informants). The third use was to sell dry hay using units of a 'backload' (20 kg) carried by women[60] at an estimated cost of KSh 200–500 per load. During the droughts of 1983/84, the market price for a backload of hay was KSh 30, compared to the 1991/92 droughts, when it was KSh 45. By 2008, the price of a backload of grass was KSh 300; in 2011, it had increased to KSh 500–600. Most (94.9%) private enclosure owners participated in hay sales.

From 2010 to the present,[61] the price of renting private enclosures has increased more than a hundred-fold. After a good wet season growth, by 2016, standing hay in a 15–20 ha enclosure was rented at between KSh 50,000 and 70,000.[62] As mentioned, some farmers with larger range enclosures harvested the hay, selling it between KSh 300 and 600 per 20 kg bale, thus making huge profits.[63] These economic opportunities have motivated the enclosing of further large areas, thereby creating radical changes in land use between the 1990s and 2000s.

The flourishing of private range enclosures reflects a breakdown of the traditional system of land use.[64] For example, the area along the frontier covering 146 km^2—accounting for about 39 percent of the total land area in the Obbu region—has been excised from communal grazing lands. There have been significant impacts on simultaneous changes in land cover.[65] The land cover maps[66] of the border region between 1986 and 2010 show that the closed grasslands (i.e., the enclosures) had increased by more than 70 percent (Fig. 5.7). By the 2000s, private range enclosures had expanded, covering an area of 600 km^2. An unfortunate outcome of the expansion of enclosures has been an increase in conflicts over land access rights.

[60] Coppock, D. L., *The Boran plateau of southern Ethiopia: synthesis of pastoral research, development and change, 1980–1991* (Addis Ababa, ILCA, 1993).

[61] In 2020/2021 when this work was being written, renting of pastures has become a widespread practice.

[62] Personal experience.

[63] Personal communication with Kadir Huqa Gulleid, Sololo-Ramata 2019.

[64] Reda, K. T., 'Dynamics in pastoral resource management and conflict in the Borana rangelands of southern Ethiopia,' *African Security Review* 25 (1): (2016): 31–43.

[65] Kamara, A. B., Swallow, B. and Kirk, M. 2004, 'Policies, interventions and institutional change in pastoral resource management of Borana, southern Ethiopia,' *Development Policy Review* 22 (4): 381–403.

[66] Roba, H. (et al.), *Pastoralist mobility in northern Kenya and southern Ethiopia* (Nairobi, 2010).

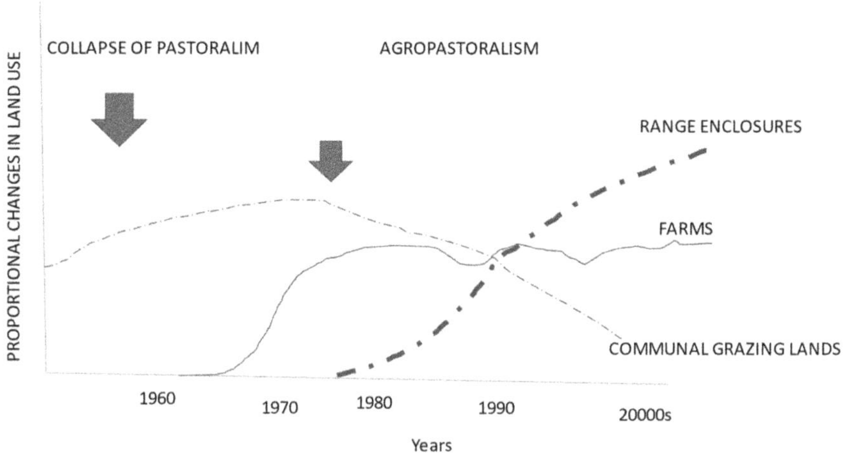

Fig. 5.7 Land cover changes along the northern Kenyan–southern Ethiopian frontier between the 1980s and 2000s

Conflicts occurred when outsiders encroached on the grazing reserve of the *de facto* right holders. Traditionally, community elders are responsible for settling disputes in intra-ethnic group conflicts.[67] Using specific bodies of rules to solve a myriad of problems,[68] the customary institutions encouraged the de facto property rights holders to fence part of the commons,[69] yet in practice, the establishment of private enclosures posed a significant threat to the authority of the customary rights holders. Under traditional resource management systems, groups used collective action to regulate land use and reinforce decisions by applying rules and regulations.[70] The collectives protected the rights of private individuals despite

[67] Degen, A.A., 'Transformation of Borana from nomadic pastoralists to agropastoralists and shifts of livestock from cattle to include more goats, camels and sheep in southern Ethiopia,' *International Journal of Business of Globalisation* 6 (3/4) (2011): 292–312.

[68] Ndunda, E. M., Mutinda, M., Murenga, H. and Bor, E., 'Contribution of collective action groups on the socio-economic wellbeing of agropastoralists in Makindu sub-County, Kenya,' *International Journal of Contemporary Applied Research* 5 (9) (2018): 199–2019.

[69] Beyene, K., 'Exploring incentives for rangelands enclosures among pastoral and agropastoral households in eastern Ethiopia,' *Global Environmental Change* 19 (2009): 484–502.

[70] Adebayo, R. M., Adeoye, A. S., Omisore, O. A. and Alabi, A. O., 'Indigenous institutions for collective action in fostering peace for sustainable land management among pastoralists and crop farmers in Ogun state, Nigeria,' *Nigerian Journal of Rural Sociology* 16 (3) (2016): 56–63.

any conflicts of interest between the commons and enclosure management.[71]

From the above analysis, we could gauge a broad set of economic relationships that highlights the extent to which the multi-faceted land-use changes influenced the growth of agropastoral production as an adaptive strategy. So, a relevant question is whether the system has retained resilience.[72]

How May We Explain the Resilience of Agropastoralism Under Changing Land-use Conditions?

In responding to the question, we are not ascribing to a single process or linkage. Various methods created unique pathways which shifted the system from one threshold to another. The pressure points in specific cases may differ, which might be described by the adaptive cycle framework[73] (see Chap. 1). The adaptive cycle starts with the exploitation phase (r) when pastoralist herds used mobility across the expansive and varied rangelands. Disasters (in our case, war) collapsed herd demography (Ω), followed by rapid transformation (reorganization) of land use (α).[74] The outcomes varied according to the pathways already described. Thus, the history of pastoral collapse and economic transformation shows one iteration of the adaptive cycle between 1963 and 2008, followed by a second iteration from 2010 to 2016 when the monetarization of land became attractive. Therefore, the exploitation, collapse, reorganization (transformation), and conservation phases can be organized chronologically (Fig. 5.8).

The 1980–1990 period was characterized by the expansion of farming and the establishment of range enclosures along the northern Kenyan–southern Ethiopian frontier. However, the severe drought of 1991/92

[71] Beyene, K., 'Exploring incentives for rangelands enclosures among pastoral and agropastoral households.'

[72] Pelling, M. and Dill, K., 'Disaster politics: Tipping points for change in the adaptation of sociopolitical regimes,' *Progress in Human Geography* 34 (1) (2010): 21–37.

[73] Holling, Crawford S. 'Resilience and stability of ecological systems,' *Annual review of ecology and systematics* 4.1 (1973): 1–23.

[74] Holling, Crawford S. "Resilience and stability of ecological systems." *Annual review of ecology and systematics* 4.1 (1973): 1–23.

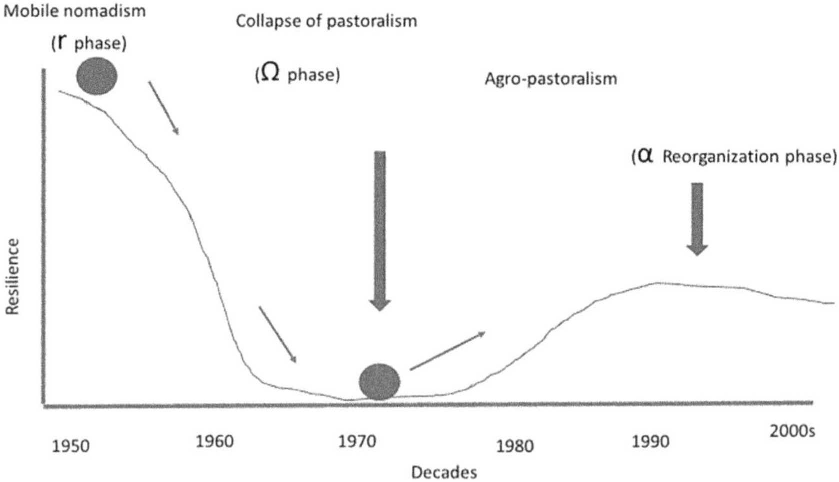

Fig. 5.8 Adaptive cycle framework illustrating pastoral socio-economic collapse and reorganization, the 1950s–2000s. The ball represents the trajectory of land use change

discouraged households from adopting new methods, forcing them to reconsider whether to focus on managing crops or stock. The 1990s–2000s might be termed the transformation phase of the adaptive model, and this period coincided with a slow and insignificant expansion of farming compared to the proliferation of private range enclosures.

The motivations for these land-use changes include—first, the households needed private pasture reserves to avoid buying hay for drought-weakened animals. Second, pasture reserves could be rented out at a price, thus improving household income. Third, there were increasing prospects for using land as a future investment. Growing markets for land near settlements had witnessed individuals selling undeveloped plots to earn substantial incomes (personal observation). Fourth, unlike private pasture reserves, farming was constrained (see earlier section on constraints), thus demotivating households to plant crops. For the Obbu households that temporarily abandoned farming, the farms left fallow were readily used for selling standing hay or renting out as fodder banks. So, enclosures motivated more individuals to claim parts of the communal grazing lands, creating momentum.[75] The rest of the community accepted the changes as a *fait accompli* due to expected land-use transformations with adverse consequences for livestock drought management triggering pathway E, described in Chap. 6. The potential application of this innovative adaptive strategy to fodder scarcity will be discussed in Chap. 10.

[75] Graham, O., *Enclosure of the East African rangelands: Contemporary trends and their impact*, ODI, Pastoral Development Network, 1988.

CHAPTER 6

The Individual in Drought Livestock Management Strategies: Mobility as a Proxy for Pastoral Resilience, the 1980s–2011

Understanding the event historical context of the analysis of this study is essential for appreciating individual herder drought management strategies. This research was motivated by a letter I received from a friend[1] in 1983 while I was conducting my graduate studies at the University of Arizona, Tucson, in the USA. The letter's author opened with the following words, 'The situation [after the 1983 drought] in the whole … [region of Obbu] is pathetic to the extent that human lives are at risk.' He reported the failure of the November–December (short) rains in 1983, followed by isolated showers that fell in various places but were insufficient to elicit growth of forage or to supply rainwater for the herds. He explained how pastoralist herds from southern Ethiopia, equally devastated by the drought, and those from the Obbu region converged on where they overgrazed the little vegetation growth that existed and exhausted the rain pools. The primary sources of water—surface dams—had dried up. The boreholes had broken down and were temporarily put out of service. This compelled the people to return with their cattle to the pre-drought settlements, where water and grazing were extremely scarce. Women traveled distances of over 20 km only to collect 20 liters of water, which they carried on their backs. By March 1984, 2500 cattle had perished. His own family had lost 90 heads of cattle. In the same letter, he briefed me about

[1] Letter from Mwalim Galma Dabasso Aga, Sololo Ramata, 1983.

G. Oba, *Pastoralist Resilience to Environmental Collapse in East Africa since 1500*, https://doi.org/10.1007/978-3-031-48291-5_6

my family herd, which I had invested significantly before I left to study in the USA.

As a graduate student at the time, I was determined to understand how individual herders responded to drought management challenges. Subsequently, I requested colleagues in Kenyan universities and colleges—who came from the same home area—to start a survey to provide a baseline for monitoring individual herders in the future.[2] I had, however, no inkling that this study would take 29 years to complete. Taking advantage of this rare data collection, we focus in this chapter on individual herd drought management strategies during six regional droughts that collapsed pastoralism. Chapter 5 mentioned how warfare curtailed herd mobility, followed by pastoral collapse. Accordingly, the return to mobility could be viewed as a sign of the recovery of pastoralism.[3] Mobility responds to the spatial and temporal variations of rainfall and pasture—between seasons and during drought years.[4] The geographical scale of herd mobility, including trans-frontier herd movements, varied according to the severity of droughts, the spatial distribution of water and grazing resources, and political security.[5] The decision to conduct mobility varied among herding units, depending on factors such as herd size, wealth, accessibility to labor, distances traveled, and social relations in the target locations.

We will elaborate on drought management histories using ten herders as case studies. This is due to the realization that drought management decision-making is seldom universal. Even when herd mobility is used as an adaptive strategy, uncertainty is a characteristic feature of herd management in drought years. Consideration needs to be given to whether management involves single herding units, cooperative herds, or splitting of herds by reproductive classes. The decisions might vary from intensive to

[2] I acknowledge the late Honorable Galma Godana Haapi and Waqo Gufu for conducting the initial survey.

[3] Turner, M. D. and Hiernaux, P., 'Changing access to labor pastures, and knowledge: The extensification of grazing management in Sudano-Sahelian West Africa,' *Human Ecology* 36 (2008): 59–80.

[4] Samuels, M. I., Allsopp, N., and Hoffman, M. T., 'Changes in pastoral mobility in a semi-arid montane region of south Africa,' *African Journal of Range & Forage Science* 38 (1) (2020): 1–13.

[5] McPeak, J.G. and Barrett, C. B., 'Differential risk exposure and stochastic poverty traps among East African pastoralists,' *American Journal of Agriculture Economics* 83 (3) (2001): 674–679.

extensive drought management.[6] We contend that oft-generalized discussions about pastoralists' drought management strategies[7] might not enlighten us about how individual herder drought management functioned at local and regional scales.

Terrence McCabe pioneered this individual approach,[8] among others, in his study of southern Turkana, proving community and personal decisions before, during, and after droughts. For the individual herder, each action (herd splitting, mobility between local landscapes, and regional migrations) involved strategic decisions to reduce any detrimental impacts on the herds.[9] Our research concerns cattle pastoralism because their drought management involves regional mobility, unlike small stock and camel pastoralism practices[10] that use local adaptive strategies.

If mobility indicates pastoral resilience, factors that disrupt herd movements would serve as proxies of pastoral collapse. At the same time, mobility—albeit at reduced distances—would imply recovery of pastoral resilience.[11] By exploiting heterogeneous landscapes, herding units could vary their responses to disaster shocks so long as rainfall variability had not induced total herd die-offs. Despite a general hypothesis that larger herding units that practiced mobility experienced lower herd losses than small herding units that practiced local drought management strategies,[12] this debate has not been concluded. We intend to enlighten the arguments based on our collected evidence in southern Ethiopia and northern Kenya herd movements.

The landscapes of the southern Ethiopian escarpment served as drought-year fallback rangelands for the Obbu Borana herds. The geopolitical distribution of natural resources influenced the direction of the

[6] Robinson, L. W. and Berkes, F., 'Applying resilience thinking to the questions of policy for pastoralist systems: Lessons from the Gabbra of Northern Kenya,' *Human Ecology* 38 (2010): 335–350.

[7] Roth, E. A., 'Traditional pastoral strategies in a modern world: An example from Northern Kenya,' *Human Organization* 55 (1996): 219–224.

[8] McCabe, J. Terrence. *Cattle bring us to our enemies: Turkana ecology, politics, and raiding in a disequilibrium system* (University of Michigan Press, 2010).

[9] Belay, K., Beyene, F. and Manig, W., 'Coping with drought among pastoral and agro-pastoral communities in Eastern Ethiopia,' *Journal of Rural Development* 28 (2005): 185–210.

[10] This is partly attributable to the abundance of browse vegetation.

[11] For processes involved see Adger, W. Neil. 'Social and ecological resilience: are they related?' *Progress in human geography* 24.3 (2000): 347–364.

[12] McPeak, J. G. and Little, P. D., 'Applying the concept of resilience to pastoralist household data,' *Pastoralism: Research, Policy & Practice* 7 (2007): 14.

transhumance,[13] as shown later. We selected herd mobility (short and long distances) within northern Kenya and the Ethiopian frontier as an indicator of adaptive drought management strategies. We used lengths traveled, herd size, herd mortality, and access to labor to evaluate the performance of pastoralist decision-making. Our case studies investigated the specific local drought management strategies households use. We were especially interested in how the entrepreneur families differ in their drought management strategies from the other homes. Finally, we examined the long-term effects of drought on households and their herds (see Chap. 7).

So, in the first section, we describe adaptive drought management strategies. The second section presents the history of the droughts of the 1960s–1990s. The third examines methods used in decision-making by cooperative herding units during the 1980s and early 1990s droughts. The fourth section discusses case studies of individual herding units over the same period, and the fifth section describes adaptive strategies during the deficiencies of the late 1990s–2000s. The sixth section uses adaptive drought management strategies to discuss pastoral resilience.

Framing Adaptive Drought Management Strategies

Adaptive drought management strategies involve individual choices that might be shared with cooperative or other family herding units.[14] If the outcomes were unfavorable, the impacts of multiple droughts posed an existential threat to the pastoral economy.[15] The slow onset of the disaster might produce false expectations among individual herders,[16] thus seriously affecting their decisions, especially during multiple droughts when

[13] Asiwaju, A. I. 'Borders and borderlands as linchpins for regional integration in Africa: Lessons of the European experience.' *Africa Development/Afrique Et Développement* (1992): 45–63.

[14] Oba, G., *Ecological factors in land use conflicts, land administration and food security in Turkana, Kenya*, London, ODI PN 32a, 1992.

[15] Mwangi, Margaret. "Diverse drought spatiotemporal trends, diverse etic-emic perceptions and knowledge: implications for adaptive capacity and resource management for Indigenous Maasai-Pastoralism in the rangelands of Kenya,' *Climate* 4.2 (2016): 22; McPeak, John G., and Christopher B. Barrett. 'Differential risk exposure and stochastic poverty traps among East African pastoralists,' *American Journal of Agricultural Economics* 83.3 (2001): 674–679.

[16] Zamani Gh. H., Gorgievski-Oujvesteijn, Zafshani K., 'Coping with drought: Towards a multi-level understanding based on conservation of resource theory,' *Human Ecology*. 10.1007/s 10745-006-90340-0.

pasture and water scarcity made drought management strategies ineffective.[17]

We will examine drought management decisions by individual herders, for whom mobility was the traditional mode of responding to grazing and water variability. There are two types of mobility—seasonal mobility and escape mobility. Seasonal mobility responds to spatial and temporal distributions of rainfall and pastures. Escape mobility is used in response to environmental constraints—scarcity of range and low nutritional quality—and is practiced during drought.[18] At the community level, as illustrated by our case studies, the benefits of mobility would depend on the duration of stress and other ecological and political factors.[19]

Herd mobility was traditionally managed by grazing associations that linked the communities within Kenya with those across the frontier in southern Ethiopia (Table 6.1). Before the collapse in the 1960s, the Obbu Borana households kept two systems of livestock management—semi-sedentary camps and satellite herds. Older people, women, children, calves, and milk herds called *hawiicha* were based in the base camps (*warra*). Before the war, the grazing orbits of the milk herds were at most 20 km from the main settlements. The satellite herds (*forra*) consisted of dry cows, bulls, and pregnant cows and were disallowed in the home-grazing rangelands during the wet season when they were mobile to access remote pastures.[20] Consequently, the proportion of milk (*hawicha*) and dry (*forra*) herds in any one place reflected the rainfall conditions and

[17] Endfield, H. H., *Climate and society in colonial Mexico: A study in vulnerability*, Blackwell, Publishing, RGS-IBG, Oxford, 2008; Natsagdorji, B., P., and Batnasan, N. 'Vulnerability of Mongolia's pastoralists to climate extremes and changes,' in Leary, N., Conde, C., Kukarni J., Nyong, A. and Pulhin, J. (Eds), *Climate change and vulnerability*, Earthscan, London, 2008.

[18] Monod, Theodore. *Introduction [à Pastoralism in Tropical Africa]* (Oxford University Press, 1975); Homewood, Katherine, Pippa Chenevix Trench, and Patti Kristjanson. "Staying Maasai? Pastoral livelihoods, diversification and the role of wildlife in development." *Staying Maasai?* Springer, New York, NY, 2009. 369–408; Little, Peter D. *The elusive granary: herder, farmer, and state in northern Kenya*. No. 73. (Cambridge University Press, 1992).

[19] See also, Nkedianye, David, et al. "Mobility and livestock mortality in communally used pastoral areas: the impact of the 2005–2006 drought on livestock mortality in Maasailand." *Pastoralism: Research, Policy, and Practice* 1.1 (2011): 1–17.

[20] Campbell, D. J., Lusch, d. P., Smuker, T.A. and Wangui, E.E., 'Multiple methods in the study of driving forces of land use and land cover change: A case study of S.E. Kajiado, District, Kenya,' *Human Ecology* 33 (2005): 763–794.

Table 6.1 Descriptions of local and regional grazing zones

Grazing zones	*Descriptions*
Golbo lowlands in Kenya	Historically, the Obbu Borana used the arid Golbo rangelands during the wet season before returning to the Obbu grazing lands during the dry season. The vast and varied landscapes are richly endowed with rapidly growing seasonal pastures on which the region's livestock converged to recuperate from the drought stress of the previous year. However, the temporary grazing and water sources did not permit extended grazing during dry years. As soon as the water sources dried up, the livestock dispersed to the dry season grazing rangelands. The Golbo rangelands are provided with earth dams and a few boreholes to supply water for grazing throughout the year.
The Obbu transitional zone	The home-based milk herds historically used the transitional zone (semi-arid). The grazing lands were traditionally grazed in transhumance cycles between the wet and dry seasons. During drought, the Obbu region lacked permanent water sources, and the local herds relied on springs and wells across the frontier. Since the sedentarization of the Obbu Borana after the late 1970s, this narrow strip of grazing lands was converted into rangeland enclosures and farmlands, thus diminishing its purpose for grazing by milk herds. All the livestock from this zone now rely on the Golbo lowlands for wet and dry year grazing.
The Badha escarpment	Across the frontier, this sub-humid zone was historically forested, endowed with springs and wells. Since the early twentieth century, Ethiopian settlers have converted it into large-scale farmlands and later (since the 1980s) settled by Borana farmers. It has plentiful water and evergreen vegetation that is attractive for drought-year grazing. However, the Badha landscapes are unsuitable for cattle husbandry due to the prevalence of disease-causing pests and poisonous plants.
The Diirre plateau	This central plateau in southern Ethiopia has offered the richest grazing lands; its diverse landscapes, perennial grasslands, and ancient *tulla* wells made the region a focal point for Borana pastoralism and continues to be so. However, recent land-use changes in settlements, peri-urban units, farms and ranches, and bush encroachment have reduced its attractiveness for pastoralism.[a]
The Ewaso Nyiro floodplain	The floodplain of the Ewaso Nyiro River, 400 km from Obbu, has continued to be a critical drought grazing zone for pastoralists' livestock from northern Kenya. The long-distance migration was a choice made by the Obbu herders.

[a]Angassa, Ayana, and Oba, G., 'Herder perceptions on impacts of range enclosures, crop farming, fire ban and bush encroachment on the rangelands of Borana, Southern Ethiopia,' *Human ecology* 36.2 (2008): 201–215

reproductivity of the herds. Under mild drought stress, mobility was across different grazing associations (*dedha*) distributed across the international frontier in southern Ethiopia, northern Kenya, and administrative borders within Kenya, including Isiolo County. These were the arrangements before the war in the 1960s (see Chap. 4). Considering that individual herders differed in their drought management strategies, the availability of sufficient labor was a crucial factor in decision-making on herd mobility.[21]

Labor Requirements

Indeed, any loss of labor would make it impossible for families to split herds into sedentary (milk) and mobile (dry) herds. Therefore, mobility over long distances would not be a priority for families with labor shortages. Collaborating households would decide to water cattle at traditional wells or boreholes operated by diesel engines. For the livestock that depended on conventional wells, sufficient labor was needed to lift water to the surface for the animals to drink. This was a back-breaking exercise that demanded the pooling of labor resources.[22] Where the households concerned lacked sufficient labor, they would be forced to maintain livestock at the borehole pump despite the prohibitive fuel costs and the risks of keeping livestock in a degraded environment.

Consequently, households (among cooperative herding units) might opt for herd mobility, while others prefer local herd management strategies. The latter methods are ideal for small herds but not selected by large herding units. Those with small herds exploited diverse landscapes to feed their stock on tree and grass litter. Women would go up steep hillsides with difficulty accessing valleys on the Ethiopian escarpment to collect dry hay to stall-feed drought-weakened animals at home. This strenuous work forced women to walk 10 to 15 km to collect 20 kg of dry grass on their backs. On alternate days, they would walk another 20 km to the boreholes to collect water for domestic use, including watering drought-weakened animals left in their care.[23]

[21] Moris, Jon R. *Directions in contemporary pastoral development.* Overseas Development Institute, Agricultural Administration Unit, 1986.

[22] Tiki, W. and Oba, G., 'Transforming labour and technology of the ancient tula wells for watering livestock in Borana, Ethiopia,' *Nomadic Peoples* 23.2 (2019): 218–240.

[23] The Obbu Borana women traditionally used baggage animals (such as donkeys and camels) to collect water. However, after sedentarization from the early 1960s to the 1970s, and following the *shifta* war of 1965–1968, they lost them. Presently, these practices have changed, as at the time of writing in 2020, the Obbu families used donkeys and motorbikes for carrying water.

Meanwhile, the men lopped evergreen tree branches and dug up root tubers with high moisture content to feed weak livestock at home. Such provisions of supplementary feed and intensive veterinary care were able to maintain animal health. However, as should be appreciated from the presentations in Chap. 5, the Obbu grazing lands are highly fragmented, and therefore, adaptive decisions for local livestock drought management strategies are diminishing. Of crucial importance was access to water sources.

Water Sources, Watering Intervals, and Grazing Resources

During droughts, forage would lose its nutritional quality, thus not meeting animals' basal metabolic demands.[24] Consequently, in their semi-starved conditions, animals used more energy than they gained from the old forage. Additionally, distances walked caused the stock to spend needed energy, thus exacerbating the deterioration of their physical condition. Where permanent water sources are limited, the watering schedules of cattle need to be varied. Watering frequency affects cattle performance, especially within the tropics, where thermal stress is high.[25] Cattle deprived of water lost as much as 17 percent of their body weight between drinking intervals.[26] Nicholson[27] investigated the three-day watering strategy used by the Borana in southern Ethiopia during droughts. He found that cattle watered every third day drank 30 percent less water in a dry season than those consumed daily. He shows that watering cattle every third day compared to daily or two-day watering intervals had no significant effects on cattle production performance and only a negligible impact on calving rates. Moreover, the three-day watering allowed animals to graze larger areas than those watered daily. We now examine these adaptive strategies considering the history of recent droughts.

[24] Nicholson, M. J. "The cost to productivity and the potential benefits of 2-and 3-day watering of Boran cattle." *Ilca Bulletin* (1986).

[25] Western D. and Frinch, V. 1986, 'Cattle and pastoralism survival and production in arid lands,' *Human Ecology* 14: 77–94.

[26] Temple, R. S. and Rey, I., Livestock population and factors affecting them. In: Nestrel, B. (ed). *Development of animals production systems*, world Animal science 2. Elsevier. Amsterdam, 1984.

[27] Nicholson, M. J. 'How water shortage can benefit pastoral society.' 4–7.

The History of Recent Droughts

In northern Kenya, drought occurs when the long rains (April–June) fail, and the following cool season (July–September) is still dry. The short rains (November–December) may also fail. This means that December through March might remain an abnormally dry season. If both the long and the short rains fail, drought occurs. Following the onset of drought, the earth dams dry up, and the pastures become exhausted. The severity of drought is weighed in terms of the scarcity of grazing and water when the livestock and people need to travel long distances to seek these resources.

The local people often assign names to individual droughts. The word *oola* is a generic term denoting failure of expected rainfall and is used with an added epithet referring to specific events that characterized a particular drought stress. According to local memories of the cyclical droughts (see Chaps. 3 and 10), the 1983/84 drought shared characteristic features with the 1940s (traditionally referred to as the 'droughts of hides and skins'—*oola qoolaaji*).[28] Like earlier droughts, cattle hides and skins were more highly prized at the markets than the drought-weakened animals during the drought of 1983/84 (*oola diima suugha*).[29]

Recovery during later years in the 1980s was progressing when the 1992 drought disrupted herd growth.[30] The long rains of 1990 were short and poorly distributed, while the short showers of 1991 were a failure. By September/October 1991, water in the earth dams had dried up. The borehole engines in Uran and Golole had broken down and remained out of operation. The 1991/92 period is also remembered because of the combined effects of drought stress and disease epidemics in cattle. The disease is transmitted by blue screw flies (*Calliphoridae* family) that laid eggs in the orifices (i.e., nose and anus) of live cattle—the maggots (i.e., the insect larvi) fed on the flesh of the living cattle, causing myiasis and death. We now apply this background to the droughts of the 1980s and 1990s, considering cooperative and individual herding units.

[28] Donaldson, T. J. *Pastoralism, and drought: A case study of the Borana of southern Borana of southern Ethiopia*, MPhil Thesis, University of Reading, 1986.

[29] USAID provided massive food aid. The yellow maize and oils were claimed to have nourished the inhabitants.

[30] Oba, G., 'The effect of multiple droughts on cattle in Obbu, Northern Kenya,' *Journal of Arid Environments* 49.2 (2001): 375–386.

Cooperative Herding Units During the Droughts of the 1980s and 1990s

In this section, we discuss the first four of our ten case studies, which focus on the adaptive strategies of cooperative herding units.

Cases 1–4: Cooperative Herding Units

We examine the strategies by cooperative herding units in which the households of Galgalo Deemo (case 1), Godana Abduba (case 2), Diima Roba (case 3), and Guyo Galgallo[31] (case 4) participated. Each of the families shared herding with different households. According to the four families, the 1983/84 drought was among the worst in decades. The households identified scarcity of food for people and scarcity of grazing and water for the stock as challenges to adaptive drought management. The primary water sources were boreholes. At the Golole borehole, supplies of fuel for the pumps were sporadic, which meant that the borehole engines pumped water for shorter intervals and—combined with frequent engine breakdowns—the livestock and people were forced to go without water for days. The individual herders described their adaptive strategies as follows.

Diima Roba, for example, reported that he left his weak animals at Uran and moved the stronger ones across the frontier. He practiced intensive herd management, lopping branches of evergreen trees to feed them. At home in Uran, he sold 20 goats to buy food for the family and veterinary drugs to treat his cattle. In total, he lost 25 heads of cattle, which represented 85 percent of his total herd. Guyo Galgallo had a similar experience, whereby despite his intensive herd care, he lost 15 of his cattle, while all his goats (20 in number) survived. He also sold cattle and goats to purchase food for the family.

These strategies might be contrasted with those of Galgallo Deemo, who used mobility in response to the isolated rainfall in 1983 in the lowlands of Obbu. Because of inadequate pastures and water supplies, he moved his herd back to the base camp in Mukutano. He also considered

[31] We have used fictious names because in the original interviews, the names of the herders were not recorded. Their drought management strategies were comparable as well as varied, with each of them reflecting different decision systems which individuals preferred, as opposed to taking cooperative decisions.

moving his herd to Gaarab Gugho in the Golbo plains, from where the livestock could walk a long distance for watering at the Walda borehole (see Fig. 2.3). When rainfall showers fell in Ogaa, he moved his stock to the area, not realizing that the showers were insufficient to grow grass. After a few weeks, Galgallo Deemo and other herders sent range scouts to assess the water and pasture situation where light rain had fallen at Adadi and Amballo. This lasted for only a week. Desperate, he returned with the remaining herd to his home base in Mukutano.

Galgallo finally decided to move his herd across the frontier into Ethiopia and camped around Tuka, where half of his cattle died. He proceeded with the remaining animals to Makanisa (still on the Ethiopian side) and camped there for several months. He moved the surviving animals to Muudi Hambo near Moyale for three months. He crossed the frontier and camped at Iladdu (on the Ethiopian side). At this place, Galgallo Deemo and his cooperative herders lost much of their stock—Galgallo lost 80 percent of his pre-drought herd. Following the arrival of showers in the Obbu area, they moved the few surviving animals back to Mukutano. Next, we compare six individual herders' strategies according to wealth levels.

Wealthy Herders

Case 5: Guracha Damballa

Guracha Damballa, lives in Mukutano. He is an acknowledged and resolute herd manager. Before the drought of 1983/84, he was considered a wealthy individual. In addition to a hotel run by one of his wives, he had gotten camels and split his cattle herds between the home milk herd and the mobile *forra* herd. Guracha Damballa used mobility to manage his milk and mobile herds like others. However, he had a bitter experience during the 1983/84 drought. He migrated with his dry mobile herd (*forra*) across the Ethiopian frontier, where his initial drought refuge was at Muudi Ambo. Meanwhile, the milking herd camped at Dukke and watered at the Walda borehole in the Obbu area. The latter herd moved to Gaarab Guddo—still in the Golbo plains of northern Kenya (see Fig. 2.3)—where a sporadic rain shower occurred in 1984. Unfortunately, the rains were insufficient to help animals recover from starvation. By June 1985, his herds moved to Moyale, settled at Kinnisa along the frontier, and watered at the wells of Holale on the Kenyan border. The herds from the region heavily stocked the area. His stock began dying. At that camp, he lost 75 percent of the 150 head of cattle in his mobile *forra* herd. The remaining animals recuperated after the drought had broken. The years 1986 to 1989 were average, during which herd recovery progressed.

In 1990, Guracha and his colleagues recognized the signs of the coming 1991/92 drought—the rains were sporadic, pasture growth was insufficient, and rainwater was exhausted, forcing the herds to cross the frontier. By this time, livestock mobility across the international border had suffered impediments due to political problems in Ethiopia.[32] During the 1991/92 drought, a livestock epidemic and scarcity of water and grazing resources caused stock conditions to deteriorate further. This was the dilemma that Guracha Dambala faced.

Before the conditions worsened, he hatched a plan to move his herd 400 km to Isiolo County[33] (Fig. 6.1) before drought stress set in. He shared his plans with other herders to avoid the calamities of the 1983/84 drought. The choice available to Guracha and his colleagues was to trek the mobile herd to the floodplain of the Ewaso Nyiro River (the Lorian swamp) in Isiolo County, where the flood-fed pasture had regenerated (Figs. 2.2 and 6.1).

While he was with the *forra* herd in the floodplain of the Ewaso Nyiro River, the conditions in Obbu deteriorated—as expected, the rains had failed. He, therefore, decided to travel back to Obbu by road to deal with the stock he had left behind. With the help of a hired laborer, he moved the milking herd and camped at Elfatu (East of Hidi Lola town) across the frontier. The area was supplied with livestock water, but environmental conditions (such as coldness and poisonous plants) threatened herd survival. Additionally, there was a foot and mouth disease outbreak among local herds. Then, there was a sporadic rainfall shower in Obbu. Unfortunately, the decision to move back was based on insufficient information. Ultimately, he was left with only three cows—all the calves had perished.

Disappointed with losing the milking herd, he returned to the Waso herds in Isiolo. By then, the floodplain pastures had been exhausted, and the livestock was moved to Yamicha, supplied by a borehole. However, the rangelands around the borehole were overgrazed, and forage was scarce.

[32] Schlee, G. and Shongollo, M. 'A Local war and its impact on ethnic and religious identification in southern Ethiopia,' *GeoJournal*, 36 (1995): 7–17.

[33] This was before the region reverted to the county administration.

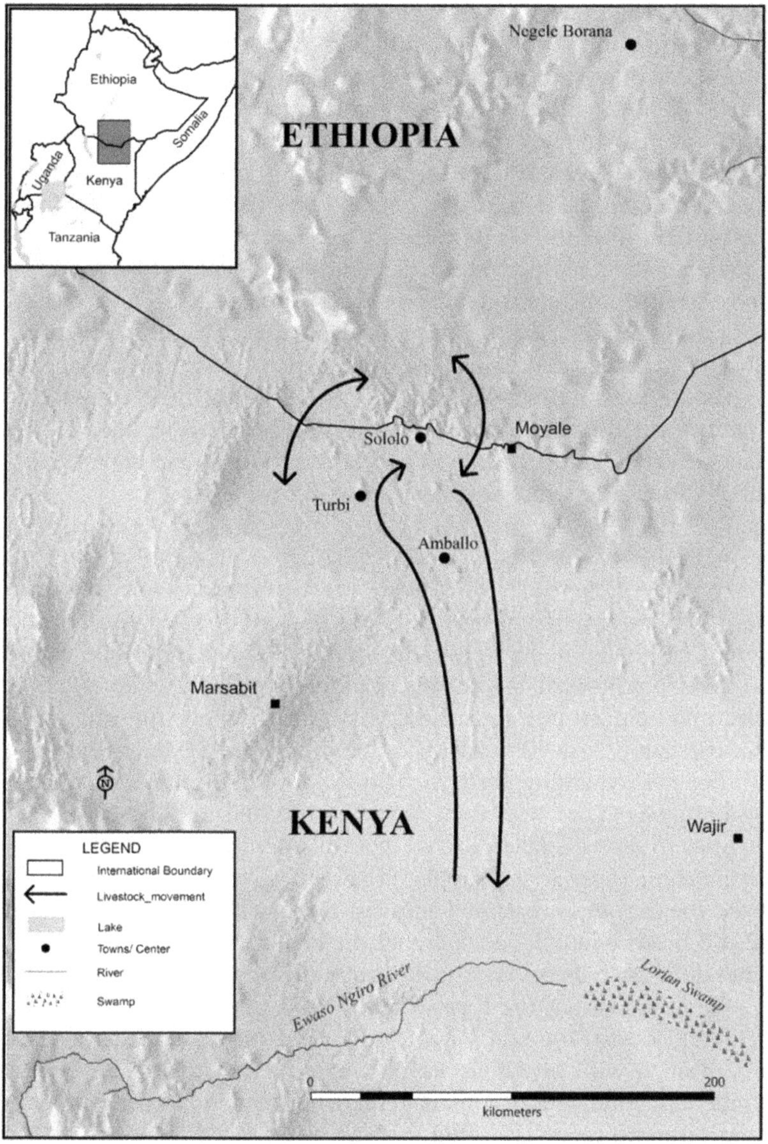

Fig. 6.1 Regional grazing movements of the case study families between 1983 and 2011

Additionally, the livestock was kept without water for two to three days. On a watering day, individual animals became overhydrated (*quuran-quuro*), collecting water in their body tissues, and as a result, he lost 21 animals. While the herd was still at Yamicha, the rains arrived. However, some drought-weakened animals got stuck in the mud, and he lost 11 more animals. Weeks later, the condition of the surviving stock had improved following the rains. He trekked the surviving herd 400 km back to the base camp in the Obbu region.

Guracha Damballa had learned lessons. He was convinced that the cause of his drought management problems was associated with the scarcity of pasture. Taking advantage of land-use changes in Obbu (described in Chap. 5), he decided to enclose a large piece of grazing land to conserve pasture for the calves during drought. Another motivating factor for this action was the loss of access to the frontier grazing due to political conflicts on the Ethiopian side of the border. We now compare Guracha Damballa's herding strategies with the sixth case family.

Case 6: Waqo Eebessee

The late Waqo Eebessee, a resident in the Uran area, was a wealthy herd owner.[34] The advantage for his family was that they had a large labor force. Different family members were assigned additional responsibilities—some cared for drought-weakened calves at home (women), while others cared for the milk and mobile *forra* herd (by men). During the early drought phase, the family lost 90 heads of cattle (both mature and immature animals). For the remaining herd, the family adopted local intensive management strategies. They sent their stock up the hillsides and bush-covered landscapes to feed on fallen leaves. They also reduced the watering intervals from three to two days, supplied their stock with veterinary drugs, and stall-fed the drought-weakened animals. By the arrival of the rains in April 1992, 80 heads of stock had survived the drought.

The difference between the households of Guracha Damballa and Waqo Eebessee is that the former practiced herd mobility over long distances. In contrast, the latter practiced local drought management strategies combined with intensive personal care of the stock. The household that practiced intensive local management succeeded in saving more cattle during drought years than the one that practiced long-distance herd movement. We now compare these cases with four other households,

[34] Waqo Ebessee had died before the 1991/92 drought.

ranked as medium wealth and poor. Two of the households practiced entrepreneur livelihoods in addition to managing cattle.

Medium-level and Poor Households

We have already discussed cases of wealthy herders, where long-distance herd mobility or intensive local management was preferred. It is interesting to understand the scope of adaptive strategies by medium wealth and poor households. We will examine the case of four herders, including a female-headed family. We interviewed the families of Tana Galgallo, Tume Galgallo Sora (a woman), Galgallo Jattani, and Guyo Qae Dabasso, who were all residents in the Uran area, on their drought management strategies during the 1991/92 drought.

Case 7: Tana Galgallo
The household of Tana Galgallo had a small herd comprising 15 heads of cattle and would, therefore, be considered poor. He practiced intensive local drought management during the 1991/92 drought when his neighbors practiced regional migrations. He increased his attention to the care of his herd by lopping trees, digging out plant roots to feed them, and providing deworming drugs. As a result of his careful management, his herd survived, apart from three cows he sold to buy food for his family and veterinary medicines to treat the rest of his herd. When the short rains (October–November) broke drought conditions, his cows and traction oxen for plowing were in good health. He planted his farm using his oxen—the largest in the Uran location.

Case 8: Tume Galgallo
Tume Galgallo is a widow, and her family is of medium wealth. Her children were young and school-going, and she had a married daughter. She managed the family cattle and small stock herds. Since she lacked labor for herd management, she relied on her son-in-law to help. When the 1991/92 drought occurred, her cattle were managed around Toroba to the west of Uran (see Fig. 2.3). The area had good pasture, but the water was scarce, forcing the herders to disperse across the Ethiopian frontier to Gamo. However, the site is too hilly, and its polluted stagnant rock pools and the vegetation dominated by poisonous plant species adversely affected stock conditions.

The herd moved from Gamo to Melbana (near Mega) to water at the ancient *Tulla* wells across the frontier (see Fig. 2.2). During the trek, two cows died, and three weak ones were abandoned. At the Melbana wells, watering remained a problem since Tume Galgallo's herd did not have extra labor to water the cattle at the deep wells. Additionally, lacking food themselves, the herders returned home with the remaining herds. Then, her son-in-law did the unexpected—he abandoned the remaining stock for a casual job in the Walda refugee camp. This forced Tume to withdraw her children from school to help herd the stock.

The preceding cases show the importance of labor and local decision-making in managing drought stresses. Since herders in the same area experienced comparable environmental risks, the difference was in their individual choices. Both local drought management strategies and long-distance migrations were used in the cases presented. The outcomes were mainly because of the timing of the decisions made. In the following case examples (cases 9 and 10), we describe management variations by two entrepreneur families.

Case 9: Galgallo Jattani

Before the drought of 1991/92, the Galgallo Jattani household was considered medium wealth. As a village carpenter, Galgallo had a lucrative source of income. As with other Uran residents, the problem for the family was scarcity for domestic and livestock use. The primary source of water was the Golole borehole. On the watering days for his cattle, he was required to buy diesel for the pump. However, the lack of diesel occasionally forced him to keep his cattle from watering longer than was necessary. Further, the area of Uran-Dida was heavily grazed during the early part of 1991. Using local drought management strategies, he cut evergreen tree branches to feed his cattle, and his wife went up the steep hills on the Ethiopian side of the frontier to collect dry hay to stall-feed the calves and other drought-weakened animals.

As the drought progressed, the family changed their strategy and moved the herds across the Ethiopian frontier, camping at Qiltipe. Unfortunately, his cattle fed on poisonous plants and died; additionally, swarms of flesh-eating 'blue flies' aggravated herd mortality. In April 1992, there was a freak rainfall shower, and herds from southern Ethiopia and the whole Obbu region converged on the area. Consequently, pasture and rainwater were exhausted sooner rather than later. Desperate, the herds returned to the home base, which aggravated livestock mortality.

By the end of April 1992, when the drought was broken and rains arrived, Galgallo Jattani had only six cows—his loss was 94 percent of his pre-drought herd. Galgallo was in a quandary: should he continue to manage his few surviving stocks, plant his farm, or use his craft of carpentry to work in the Walda refugee camp? He chose the latter. His goal was to use the income to buy stock to replace the ones he had lost and purchase seeds for planting when the rains returned—the next household combined local and long-distance mobility.

Case 10: Guyo Qae Dabasso

Despite his enterprising pre-drought economy, Guyo Qae Dabasso's household performed poorly. During the 1983/84 drought, he had gone to Meru in Kenya to seek employment as a farm hand and later as a security guard in Nairobi. Over the years, he saved enough money to replace the animals he lost and returned home to experiment with the retail business. Guyo Qae used his strategic location near the Ethiopian border to conduct trade. His family sold milk to Ethiopian farmers across the frontier and bought food from Ethiopian markets. Later, he participated in the stock trade across the border and bought grains, which he sold to the residents in Uran on the Kenyan side of the border.

During the 1991/92 drought outbreak, Guyo Qae had 22 cattle and a similar number of sheep and goats. Additionally, he had a small retail business and a farm. Following the failure of the long rains (April–June) in 1991, Guyo Qae called a meeting of his family members (*maala maari*) to decide how to manage their herds. He attempted intensive herd care as the drought progressed by providing tree branches and dry hay as supplementary feed. His family realized that this strategy did not sustain the herd. He, therefore, decided to cross the frontier with his cattle. He initially settled at Teeso Alle on the Ethiopian side of the border. After a week at the place, the herd's health improved, and the weak animals recuperated. However, the chilly weather of the Ethiopian escarpment, combined with a high worm load, caused the condition of his herd to deteriorate. Guyo Qae provided anthelmintic drugs to treat his herd. Yet the rest of the herd began to succumb to the combined stress of disease and the worsening drought. He moved the herd to Erdar to water at the ancient *Tulla* wells. Here, he lost two more cows. Guyo Qae decided to move on to Tumtu, still on the Ethiopian side of the frontier. He lost two calves on the way, and seven more cows died after arrival. The area was infested with ticks and livestock diseases. He was desperate. He returned to his Uran home camp, losing 97 percent of his herd. The example of Guyo Qae demonstrates crisis management in drought adaptive strategies. He responded by purchasing cheap food at the refugee camp, selling it to the locals, and attempting to revive his livestock trade.

During the post1990s droughts, the Obbu Borana pastoralists continued to experience cyclical droughts into the 2000s. We now summarize their adaptive strategies using anecdotal information from herders.

The Droughts of the 1990s–2000s

The droughts of the 1990s and 2000s occurred when the traditional drought refuge areas across the Ethiopian frontier were lost due to conflicts following the fall of the military government of Mengistu Haile Mariam.[35] Changes in the resource base and political insecurities across the international border posed the greatest threats to drought management. Due to the expansion of agropastoralism and the enclosure of communal grazing lands within the Obbu region, the resource base for local drought management previously practiced for small herds had remained the same. Additionally, political conflicts along the frontier implied that long-distance mobility across southern Ethiopia had ceased. All families, including those with small herds, focused on the dry *forra* herd management model in the Golbo lowlands (Figs. 2.3 and 6.1). Additionally, small herd owners increasingly participated in cooperative herding units to consolidate labor for herding.

The year 1994 was wet, but 1995 was dry. Guracha Damballa (one of our case herders) was again caught in a quandary with limited options. This time, he needed to care for his camels, which forced him to adopt a local drought management strategy using the borehole at Walda. His cattle were watered at three-day intervals. He located his herd 50 km from the water source, where pasture was available. On a watering day, the herd trekked to the borehole, stayed for the night in the second camp, and was taken to water the next day. Following the arrival of the long rains in early April 1996, the surviving herd had recovered from drought stress.

The year 1997 was an El Niño year. The heaviest rainfall was associated with a malaria pandemic and Rift Valley fever that caused human and

[35] Schlee, G. and Shongollo, M., 'A Local war and its impact on ethnic and religious identification in southern Ethiopia.'

livestock fatalities.[36] Conditions had begun to change by 1998, when the short rains were inadequate, deteriorating further in 1999. The year 2000 experienced the most severe drought over the preceding 37 years, while in the year 2002—the short rains failed, while the long rains caused devastating floods.

The rainfall failure in 2004 further aggravated livestock mortalities.[37] Our case study families attempted their old strategies, while most households had abandoned long-distance stock migrations. For example, during the 2006 drought, Guracha Damballa tried to cross the frontier. He used his clan networks on the Ethiopian side of the border to gain access to the government ranch that served as a drought-year reserve. During the 2009 drought, he practiced local drought management strategies, while during the 2011 drought, with six other herders, he once again resumed the long-distance migration to the Uaso Nyiro floodplain pastures in Isiolo County. He again lost most of his cattle and concentrated on managing his camels and small stock herds.

The case households we have described are representative of the status and practice of drought management across the region in Obbu. Society responded in two significant ways. The first was to expand grazing into the Golbo areas that had historically been abandoned. Second, the Obbu Borana revived their indigenous resource management to adjust as part of the emerging adaptive strategies for restoring pastoral resilience.

The Restoration of Pastoral Resilience Explained

The case studies allow us to contextualize drought adaptive strategies at individual herder levels. Regarding geographical scope, we have identified two types of adaptive drought management strategies—local methods often practiced by small herd owners and regional strategies involving long-distance mobility, usually practiced by more wealthy herders. The geographical reach and the cycles of herd mobility between two or more drought events show differences among our case households. Thus, contrary to a general belief that families with larger herds experienced less livestock mortality, the case studies show more survival among smaller

[36] Little, P. D., Mahmoud, H. and Coppock, D. L., 'When deserts flood: Risk management and climatic processes among East African pastoralists.'

[37] Oluwafemi, B. A., *Vulnerability assessment of Maasai pastoralists under changing climatic condition*, 2016.

herds that practiced local intensive drought management strategies.[38] These adaptive strategies have since been adversely affected by the fragmentation of grazing landscapes in the Obbu transitional zone (see Chap. 5). We briefly sum up the changes.

Challenges and Transformation

Following the war and the collapse of pastoralism, herd mobility was abandoned until two decades later (Chap. 5). Yet, the expansion of mobility still suffered setbacks due to conflicts along the frontier and in the remote Golbo rangelands, where pastoralists from northeastern areas competed for grazing with the Obbu Borana. Under these transformed systems of land use, pastoral resilience as a process reflects complex changes (Fig. 6.2). Let us examine the processes behind the restoration of pastoral resilience.

The Obbu Borana responded to various challenges, including a shrinking resource base, competition over the remaining grazing lands, and dysfunctional institutions to build pastoral resilience. Considering that the processes were linked to environmental conditions, social relations, and

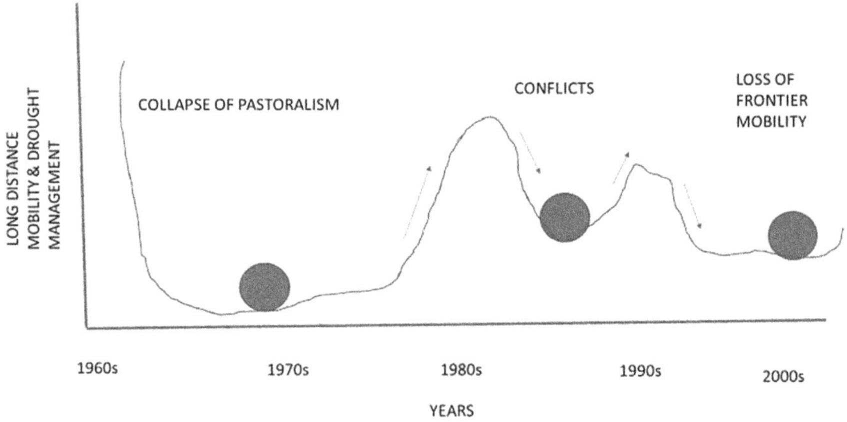

Fig. 6.2 Trajectory of drought management strategies over four decades

[38] Barton, D., Meadows, N. and Morton, J, *Drought losses, pastoral saving, and banking: A review*, DFID Advisory and Support Services Commission Project 2W0027 NR Project L0114/2001.

group cooperation, we argue that although new methods might have ensured the success of individual drought adaptive strategies, the restoration of herd mobility did not implicitly improve herd survival. However, it might have served as the building block for the resilience infrastructure.[39] Nevertheless, today, incentives to continue with traditional methods have disappeared.[40] The diminishing herds, a shortage of herding labor, and shrinking grazing lands are among the factors that have forced changes. Institutionally, traditional grazing associations have collapsed. Consequently, the society is attempting to build a hybrid institution using cooperative formal environmental conservation approaches, with aspects of the indigenous associations for managing grazing lands.

The processes are described in Chap. 5, more specifically, Fig. 5.1 E. We have already explained the changes and transformation of resource tenure. Seeking alternative land-use modes and resource tenure changes implied that the society used new adaptive mechanisms to respond to the changes.[41] The grazing lands traditionally allocated to the mobile *forra* herds had declined by 30 percent, while by the 2000s, the home-grazing lands allocated to the milk herds had disappeared (Fig. 6.3).

Another reason for changes in herd management was the availability of labor (Fig. 6.4). The problem was not the scarcity of labor but the unattractive cost thereof (KSh 300 or USD 4.2 per month in 1992). Conversely, during the droughts of the 2000s, more households (58%) relied on hired herding labor. By this latter date, on average, salaries paid to herd laborers had increased to KSh 4000 (USD 40.0 per month). Child herding labor was not in evidence, as the children attended school (Table 6.2).

Additionally, following the changes in land tenure, the Obbu Borana responded to these challenges by transforming the indigenous *dedha* council (responsible for the grazing associations) to build a hybrid institution for natural resource management, which they called *mazingira*

[39] Adger, W. N., 'Social and ecological resilience: Are they related?' *Progress in Human Geography* 24 (3) (2000): 347–364.

[40] Rasmussen, Von, L. and Reenburg, A., 'Collapse and recovery in Sahelian agro-pastoral systems: Rethinking trajectories of change' *Ecology and Society* 17 (1) (2012): 14. 105751/ES-04614-170114.

[41] Parker, D. C., Manson, S. M., Janssen, M. A., Hoffman, J. and Deadman, P., 'Multi-agent systems for the simulation of land use and land cover change: A review,' *Annals of Associations of American Geographers* 93 (2) (2003): 314–337.

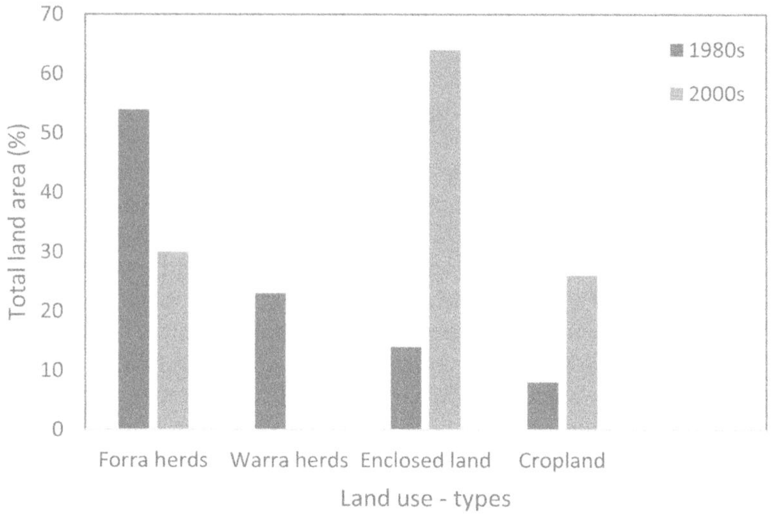

Fig. 6.3 Changes in land use allocated to forra and warra (milk) herds and croplands between the 1980s and 2000s

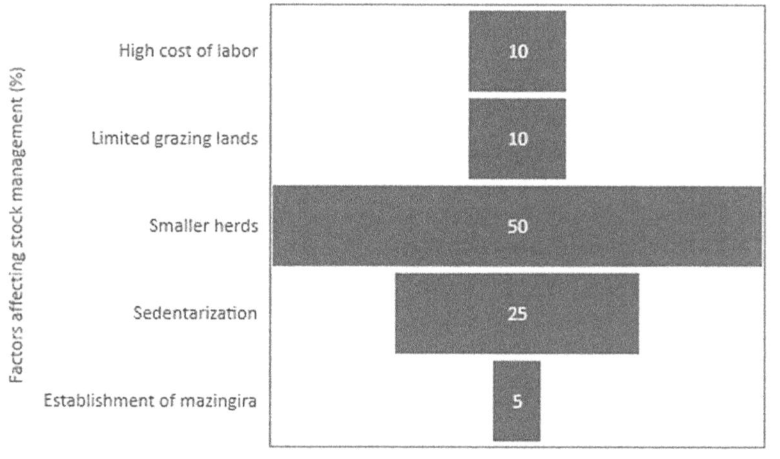

Fig. 6.4 Herder's perceptions (%) of factors affecting stock management practices

Table 6.2 Factors contributing to cooperative herding among the Obbu households

Reasons for cooperative herding	*Responses (%)*
Small herds per household	20
Children attending school	15.6
Shortage of herding labor	20
Families involved in diverse activities	28.9
Labor migration	8.9
Increased poverty	2.2

(Fig. 6.4)—a Swahili term for environmental protection—which has become a new institution.[42]

The reasons for modifying the indigenous range management system included the government administration's perception that local communities had legal powers to protect and conserve natural resources. The Borana successfully used this modification to negotiate with other settlements to access the *mazingira* grazing areas preserved during the wet season for dry season grazing. By using this formal means of environmental protection, the Obbu Borana can now regulate the use of these pastures as a form of drought reserve. The changes had, however, not minimized the mortalities of the herds. Chapter 7 presents long-term trends in cattle populations and household wealth dynamics in response to decadal droughts.

[42] New trading and administrative settlements have since been established across the Golbo low plains (Fig. 2.4).

CHAPTER 7

Impacts of Decadal Droughts on Cattle Populations: Tracking Household Wealth Dynamics, 1982–2011

This chapter is concerned with two important themes. First, it investigates the impacts of decadal droughts on cattle demography. Second, it tracks trends of household wealth in terms of cattle ownership. The key driver for both themes is rainfall variability.[1] Extended dry conditions might cause minor stress on the cattle population and corresponding household wealth, provided the next rains come on time.[2] Conversely, one or two-year droughts induce high herd die-offs, thus adversely affecting household wealth.[3] The time between the return of drought conditions is crucial for herd recovery.[4] In our analysis, we followed the traditional criterion that uses cattle to gauge wealth related to animals per capita or total cattle

[1] Gies, L., Agusdinata, D. B. and Merwade, V., 'Drought adaptation policy development and assessment in East Africa using hydrologic and system dynamic modelling,' *Natural Hazards* 74 (2014): 789–813.

[2] Ochieng, R., *Use of seasonal climate forecast and drought effects on livestock assets on Baringo county, Kenya*, PhD Thesis, Egerton University, Nakuru, 2018.

[3] Lesnoff, M., Cornianx, C. and Hiernaux, P., 'Sensitivity analysis of the recovery dynamics of a cattle population following drought in the Sahel region,' *Ecological Modelling* 232 (2012): 28–39.

[4] Coppock, D. L., Gebru, G. and Desta, S., 'Are cattle die-offs predictable on the Borana plateau?' Research Brief, PARIMA, 2008.

G. Oba, *Pastoralist Resilience to Environmental Collapse in East Africa since 1500*, https://doi.org/10.1007/978-3-031-48291-5_7

holdings.[5] Whereas considerable variations exist in household wealth,[6] it is estimated that a pastoralist requires an equivalent of 4.5 cattle per capita to meet minimum livelihood requirements.[7]

Different viewpoints have nonetheless emerged between those who associate stock populations with human population variations and those who reject this hypothesis.[8] Among the proponents, Bollig and Lang[9] studied the Pokot population, which showed a downward spiral in wealth over four decades. In southern Ethiopia, Coppock[10] attributed the decline in livestock wealth to progressive human population growth. Michael O'Leary,[11] studying the Rendille, predicted that the human population was increasing faster than that of livestock, suggesting that a progressive increase in the human population was followed by a decline in livestock per capita. The alternative hypothesis suggests that wealth hierarchies are

[5] Grandin, Barbara, 'The importance of wealth effects on pastoral production: a rapid method for wealth ranking,' In: *Pastoral systems research in sub-saharan Africa* (Addis Ababa: ILCA, 1983), pp. 237–256.

[6] Little, Peter D., 'Social differentiation and pastoralist sedentarization in northern Kenya,' *Africa* 55.3 (1985): 243–261.

[7] See Pratt, D. J. and Gwynne, M. D. 1977 'Rangeland Management and Ecology in East Africa' Hodder and Stoughton, London.

[8] In East Africa wealth hierarchies among different pastoral groups have been discussed by Sieff, D. F. 'The effects of wealth on livestock dynamics among the Datoga pastoralists of Tanzania' (*Agricultural systems* 59, 1999: 1–25); Dahl, G. 1979 '*Suffering grass: subsistence and society of Waso Borana*. Stockholm Studies,' Grandin, B. E. 1983. 'The importance of wealth effects on pastoral production: A rapid method for wealth ranking.' In *pastoral systems research in sub-Saharan Africa. The proceedings of workshop held at ILCA, Addis Ababa, Ethiopia 21 to 24 March, 1983* pp. 237–254. In West Africa, see, Sutter, D. F. 1987 'Cattle and inequality: herd size differences and pastoral production among the Fulani of north-eastern Senegal' (*Africa* 57, 198–218).

[9] The inverse relationships between human demography and livestock per capita has been reported by Bolling, M. and Lang, H. 1999 'Demographic growth and resource exploitation in two pastoral communities' (*Nomadic Peoples* 3, 16–34).

[10] Coppock, D. L. '*The Borana plateau of southern Ethiopia: synthesis of pastoral research, development, and range, 1981–91. ILCA # 5, Addis Ababa Ethiopia*', 1994.

[11] O'Leary, M. 'Changing responses to drought in Northern Kenya: The Rendille and the Gabbra. In Baxter PTW and Hogg, R. (Eds). *Property, poverty, and people: changing rights in property and problems of pastoral development*' Manchester, UK. Department of Social Anthropology and International Development Centre, University of Manchester, 1990, pp. 55–79.

highly transient, regardless of human population numbers.[12] We used household life histories among the Obbu Borana of northern Kenya to map wealth dynamics in response to disasters.[13]

The chapter is in four sections. The first section presents a framework for analyzing cattle demographics and household wealth dynamics. The second section examines factors responsible for regional and local cattle demography dynamics. In the third section, we characterize the impacts of cattle population dynamics on household wealth trends. In the fourth section, we discuss the outcomes regarding the resilience of cattle demographics and household wealth dynamics.

Framework for Cattle Demographics and Wealth Resilience

The concept of pastoral resilience implies multi-faceted processes and contexts.[14] The complex interaction between climatic variability and herd dynamics results in herd collapse and recovery cycles. The changes over time have a direct influence on household wealth dynamics.[15] In a resilient pastoral system, droughts induce herd demography and household wealth into collapse, followed by recovery.[16] The challenge is investigating how collapse and recovery in herd demographics and wealth changes may be measured.[17] Resilience reflects the capacity of herds and household wealth

[12] See Grandin, Barbara E. 'Livestock transactions data collection.' (1983): See also, Gufu, B. S., 'Sustainable pastoral livelihoods at a crossroads: Broken system resilience and institutional disenfranchisement among the Borana Oromo of southern Ethiopia,' in Muchie, M., Gumede, V., Oloruntoba, S., and Check, N. A. (eds.) *Regenerating Africa: Bringing African solutions to African problems*, Africa Institute of South Africa, 2017.

[13] See also, Gufu, B. S., 'Sustainable pastoral livelihoods at a crossroads: Broken system resilience and institutional disenfranchisement among the Borana Oromo of southern Ethiopia,' in Muchie, M., Gumede, V., Oloruntoba, S., and Check, N. A. (eds.) *Regenerating Africa: Bringing African solutions to African problems*, Africa Institute of South Africa, 2017.

[14] Koraka, S. and Little, P. D., 'Introduction: Rethinking resilience in the context of East African pastoralism,' *Nomadic Peoples* 25 (2) (2021): 165–180, https://doi.org/10.3197/np.2021.250201.

[15] Robinson, L.W. and Berkes, F., 'Applying resilience thinking to the questions of policy for pastoralist systems: Lessons from the Gabra of Northern Kenya,' *Human Ecology* 38 (2010): 335–350.

[16] Adger, W. N., 'Social and ecological resilience: Are they related?' *Progress in Human Geography* 24 (3) (2000): 347–364.

[17] Cabell, J. F. and Oelofse, M., 'An indicator framework for assessing agroecosystem resilience,' *Ecology & Society* 17 (1) (2012): 18, https://doi.org/10.5751/ES-04666-170118.

to recover from the impacts of disasters. At the same time, a collapse depicts the failure of recovery (Fig. 7.1A and B). Recovery levels vary with total herd losses and the return time of the drought. The fluctuating demographic herd structures establish theoretical benchmarks for gauging the subsequent recovery phase.[18] Our subjects of analysis are the time intervals between the return of disasters, climate variability, agroclimatic zones, and land-use intensification.

Normal herd population fluctuations are related to natural mortalities and recruitments.[19] More significant mortalities of calves and reproductive females are associated with herd collapse.[20] The relationship might be harmful if the annual mean rainfall is lower than the mean long-term rainfall.[21] Herd dynamics have direct effects on the wealth of herding households.

From Fred Bath's[22] seminal social organization and production model, a society maintains unequal regenerative economic wealth. This is pronounced because the pastoral economy is susceptible to catastrophic losses during droughts.[23] Accordingly, cattle numbers provide financial capital

[18] Rosen, S., Murphy, K. M. and Scheinman, J. A., 'Cattle cycles' *Journal of Political Economy* 102 (3) (1994): 468–492.

[19] Oba, G. 'The effect of multiple droughts on cattle in Obbu, Northern Kenya,' *Journal of Arid Environments* 49.2 (2001): 375–386; Owen-Smith, N., Mason, D. R. and Ogutu, J. O., 'correlates of survival rates for 10 African ungulate populations: Density, rainfall and predation,' *Journal of Animal Ecology* 74 (2005): 774–788.

[20] Angassa, A. and Oba, G., 'Relating long-term rainfall variability and cattle population dynamics in communal rangelands and a government ranch in southern Ethiopia.'

[21] Alemayehu, K. and Fantahun, J., 'The effect of climate change on ruminant livestock population dynamics in Ethiopia,' *Livestock Research for Rural Development* 24 (10) (2012): 1–10; Lunde, T. M. and Lindtjørn, B., 'Cattle and climate in Africa: How climate variability has influenced national cattle holdings from 1961–2008,' *Peer J.* https://doi.org/10.7717/peerJ-55.

[22] Barth, F., Models of social organization. Occasional Paper, Royal anthropology Institute 23 London: Royal Anthropological Society.

[23] Mulder, M. B., Fazzio, I., Irons, W., McElreath, R. L., Bowles, S., Bell, A., Hertz, T. and Hazzah, L., 'Pastoralism and wealth inequality: Revisiting an old question,' *Current Anthropology* 51 (1) (2010): 35–48.

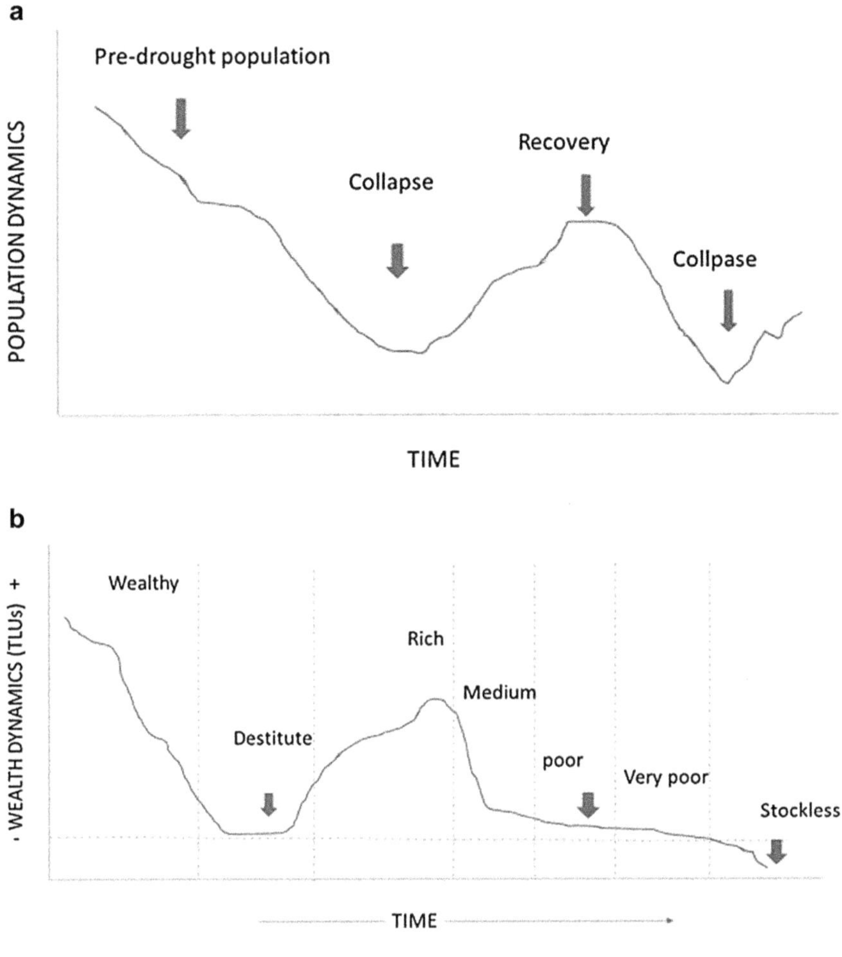

Fig. 7.1 (A) A hypothetical cattle population dynamic in response to multiple droughts and (B) household life histories of wealth ranks in response to multiple droughts

and social insurance.[24] Indeed, according to Rubel,[25] shifts in wealth among pastoralists have social security implications.[26]

The most common method of measuring household wealth is to relate livestock units (such as Tropical Livestock Units—TLUs) to the ratio of human dependents per household.[27] The TLU positively predicts the consumptive and social needs of individual families.[28] It has been hypothesized that households below the minimum figure (4.5 TLU per person) will likely drop out of pastoralism.[29] Almost invariably, herd sizes seldom remain constant, shifting between recovery and collapse.[30] As a result of multiple droughts, the pre-drought cattle holdings of individual households shifted between pre- and post-drought numbers.[31] Collecting the life histories of individual households—in terms of stock owned per unit of time is, therefore, one way of gauging wealth shifts in response to multiple droughts (Fig. 7.1B). We will apply these propositions to understand cattle demographics and wealth dynamics in response to decadal droughts. We begin by analyzing the dynamics of cattle demographics.

[24] Doran, M. H., Low, A. R., and Kemp, R. L., 'Cattle as a store of wealth in Swaziland: Implications for livestock development and overgrazing in Eastern and Southern Africa,' *American Journal of Agricultural Economics* 61 (1) (1979): 41–47.

[25] Rubel, P. G., 'Herd composition and structure: On building models of nomadic pastoral societies,' *Man* 4 (2) (1969): 268–273.

[26] Santos, P. and Barrett, C. B., 'Heterogeneous wealth dynamics: On the roles of risk and ability,' in Barrett, C. B., Carter, M. R. and Chavas, J. P. (eds.) *The economics of poverty traps* (Chicago: University of Chicago Press, 2018).

[27] BurnSilver, S., 'Presenting wealth in a changing pastoral economy: A comparison of traditional and new metrics in Maasailand, Kenya,' *Nomadic Peoples* 20 (2016): 8–34.

[28] Sellen, D. W., 'Nutritional consequences of wealth differentials in East African pastoralists: The case of the Datoga of Northern Tanzania' *Human Ecology* 31 (4) (2003): 529–565.

[29] Moritz, M., Buffington, A., Yoak, A. J., Hamilton, I.M., and Garabed, R., 'No magic number.'

[30] Tache B., *Pastoralism under stress: Resources, institutions, and poverty among the Borana Oromo in southern Ethiopia*, PhD Thesis, Department of International Environment and Development Studies, Norwegian University of Life Science Ås, Norway; Yosef, T., Mengistu, U., Solomon, A., Mohammed, Y. K. and Kefelgn, K., 'Camel and cattle population dynamics and livelihood diversification as a response to climate change in pastoral areas of Ethiopia' Livestock *Research for Rural Development* 25 (9) (2013): 1–13.

[31] Scoones, I., 'Coping with drought: Responses of herders and livestock in contrasting savanna environments in southern Zimbabwe,' *Human Ecology* 20 (3) (1992): 293–314; Little, P. D., McPeak, J., Barrett, C. B. and Kristjanseon, P, 'Challenging orthodoxies: Understanding poverty in pastoral areas of East Africa,' *Development and Change* 39 (4) (2008): 587–611.

The Dynamics of Cattle Demographics

Cattle herd dynamics may be attributed to various factors, including reproductive rates during average years, mortalities during periods of droughts, and offtake (not considered in the current study). Herd mortalities are a significant factor in the dynamics of herd demographics.[32] We begin by discussing herd mortalities.

Herd Mortalities

The cattle population data shows large fluctuations in herd sizes from one year to another. During approximately a three-decade monitoring period, household cattle demography was punctuated by five significant droughts. Population crashes occurred during the 1983/84, 1991/92,[33] and 2011 droughts. Other crashes occurred during the 2000–2002 and 2004–2005 periods (Fig. 7.2). Different stress events left their signature on structural changes in cattle populations. Reproductive females had more elasticity for surviving droughts than calves and immatures. This means that the ratio of calves to females and immatures to females provides a good picture of future herd growth.[34]

We assume that herd mortalities occurred all the time—with a 2 to 5 percent loss during an average year, while drought-induced mortalities significantly impacted cattle demography.[35] The herd crashes were associated with nutritional deficits.[36] The calamitous decline during the 1991/92 drought was attributable to disease epidemics and starvation.[37] Coppock[38]

[32] Mapiye, C., Chimonyo, M. and Dzama, K., 'Seasonal dynamics, production potential and efficiency of cattle in the sweet and sour communal rangelands in South Africa,' *Journal of Arid Environments* 73 (2009): 529–536.

[33] Note, 1990 was a favorable year, with drought-breaking out in the second quarter of 1991.

[34] DeCesare, N. As J., Hebblewhite, M., Bradley, M., Smith, K. G., Hervieux, D. and Neufeld, L., 'Estimating ungulate recruitment and growth rates using age ratios,' *The Journal of Wildlife Management* 76 (1) (2012): 144–153.

[35] Degen, A. A., 'Transformation of Borana from nomadic pastoralists to agropastoralists and shift of livestock from cattle to include more goats, camels and sheep in southern Ethiopia,' *International Journal of Business and Globalisation* 6 (3/4) (2011): 292–311.

[36] Ellis, James E., and David M. Swift. "Stability of African pastoral ecosystems: alternate paradigms and implications for development." *Rangeland Ecology & Management/Journal of Range Management Archives* 41.6 (1988): 450–459.

[37] Oba, G., 'The effect of multiple droughts on cattle in Obbu, Northern Kenya,' *Journal of Arid Environments* 49 (2001): 375–386.

[38] Coppock, D. L., 'Constraints to development of extensive livestock systems: experiences from southern Ethiopia.' *BSAP Occasional Publication* 16 (1993): 87–92.

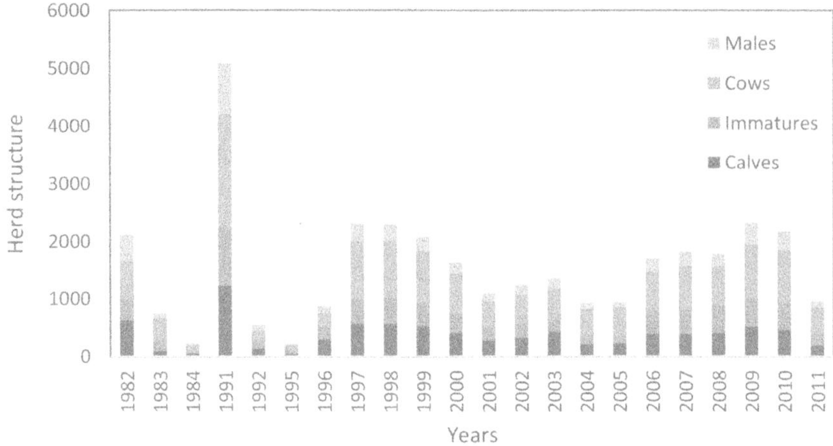

Fig. 7.2 Regional cattle herd structure dynamics from 1982 to 2011

reports cattle losses among the Borana of southern Ethiopia during the drought of 1983/84, ranging from 40 to 90 percent of immatures, 45 percent of mature cows, and 22 percent of mature males. In the same region of southern Ethiopia, Donaldson[39] reports that, on average, 55 percent of the mortality was contributed by calves compared to mature cattle, with an average mortality of 45 percent. High calf mortality has been attributed to a lack of milk.[40] Drought-induced mortalities occurred in phases.[41] By 9–12 years, most breeding females are likely to have died.[42] Patterns of herd growth and die-offs clearly show age-related mortality. Old, sick, and young animals were most vulnerable during the first phase. In the second phase, lactating cows would be affected. In the third phase, the mortality of lactating cows exposed the calves to greater vulnerability. In the fourth phase, immature and mature males might succumb. And finally, in the fifth phase, the non-lactating cows would fall victim. The trigger is climate variability, which we discuss next.

[39] Donaldson, Timothy J. *Pastoralism, and drought. A case study of the Borana of southern Ethiopia*. Diss. 1986.

[40] Coppock, D. L., 'Constraints to development of extensive livestock systems'.

[41] Homewood, K. and Lewis, J., 'Impact of drought on pastoral livestock in Baringo, Kenya, 1983–85,' *Journal of Applied Ecology* 24 (2) (1987): 615–631.

[42] Richard Trevor. *Livestock production in central Mali: Long-term studies on cattle and small ruminants in the agropastoral system*. Vol. 14. ILRI (aka ILCA and ILRAD), 1986.

Impacts of Climate Variability

Positive correlations exist between mean regional cattle populations and rainfall deviations from the long-term mean. Across agroclimatic zones (very arid, arid, and semi-arid), rainfall varies in space and time, influencing cattle production, die-offs, and recovery.[43] The impact of rainfall variability was mediated through livestock feed quality.[44] Additionally, the negative effect of the El-Niño events in 1997 was associated with disease epidemics.[45] Out of the four types of rainfall in the regression model (Table 7.1), the long rains show the highest level of correlation with the mean annual total TLU.[46] This means that the relationship between cattle holdings and rainfall deviations from the long-term mean was indirect.

Yet, rainfall alone cannot explain changes such as 1991/92. Instead, cattle mortalities were a combination of declining rainfall and disease epidemics. Rain was below average in 6 out of 12 years (Fig. 7.3). Collapses in cattle populations followed the decline in rainfall.[47]

Table 7.1 Bivariate regressions (r^2) between rainfall variability and mean total cattle (TLUs) and mean calf populations

Rainfall variability	*Mean total cattle population (TLU)*	*Mean calf population*
Short rains	0.066**	0.04**
Long rains	0.078***	0.041**
1-year lag short rains	0.045***	0.041**
1-year lag long rains	0.025^{NS}	0.036^{NS}

** $p < 005$, *** $p < 001$, Ns, not significant

[43] Little, P. D., McPeak, J., Barrett, C. B. and Kristjanson, P., 'Challenging orthodoxies: Understanding poverty in pastoral areas of East Africa,' *Development and Change* 39 39 (4) (2008): 587–611.

[44] Oba, G., *Mobility, and the sustainability of the pastoral production system in Africa.*

[45] Ahmed H., Tessema, Z., Tolera, A. and Korecha, D., 'Inter-connection between El-Niño Southern Oscillation induced rainfall variability, livestock population dynamics and pastoralists adaptive strategies in Eastern Ethiopia,' *Journal of Environment and Earth Sciences* 7 (3) (2017): 11–24.

[46] Homewood, K. and Lewis, J., 'Impact of drought on pastoral livestock in Baringo, Kenya, 1983–85.'

[47] Tuffa, S. and Treydte, A. C., 'Modeling Boran cattle populations under climate change and varying capacity,' *Ecological Modelling* 352 (2017): 113–127.

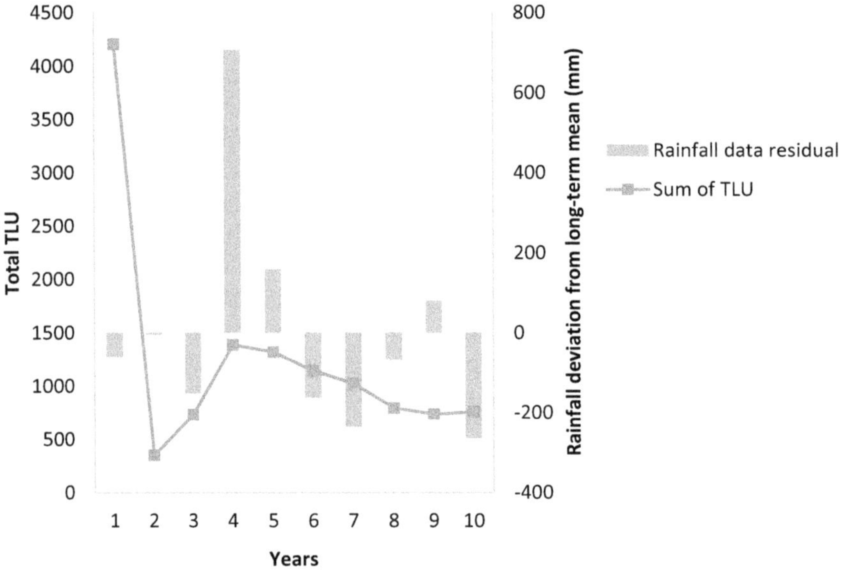

Fig. 7.3 Relationship between total cattle population (in TLUs) and rainfall deviation from the long-term mean (1991–2003)

Taking advantage of the spatial distribution of rainfall and forage,[48] the stock gained weight rapidly through compensatory body mass growth. This was also the time when mating frequencies by bulls increased. An exception was during the El-Niño events in 1997, which was associated with outbreaks of Rift Valley Fever that aggravated cattle mortalities despite excessive rainfall and high forage production.[49] The patterns of herd mortality varied across agroclimatic zones and at local scales.

[48] McCabe, J. T. and Ellis, J. E., 'Beating the odds in arid Africa,' *Natural History* 1 (87) (2003): 33–40.

[49] Jost, Christine C., Serge Nzietchueng, Simon Kihu, Bernard Bett, George Njogu, Emmanuel S. Swai, and Jeffrey C. Mariner. 'Epidemiological assessment of the Rift Valley fever outbreak in Kenya and Tanzania in 2006 and 2007,' *The American journal of tropical medicine and hygiene* 83 (2) (2010): 65.

Herd Population Dynamics Related to Agroclimatic Zones and Local Scales

Due to rainfall variability, households across the three agroclimatic zones (very arid, arid, and semi-arid) were interdependent regarding grazing lands. The herds moved across the agroclimatic zones for grazing and watering. During drought years, demographic collapses were comparable across the zones.

At local scales, we used differences in total cattle populations between drought and non-drought years to investigate the trajectory of population changes (+/–%) by sites. We identified three population patterns—related to the dynamics of mature females and males, calves, and the immatures. In total cattle populations, mortalities depicted negative herd growth (–%), while recovery caused positive changes (+%).

In the first population pattern, mortalities of reproductive females and calves during the 1991/92 drought exceeded those of other cattle age/sex classes (Fig. 7.4 A and B) at the sites of Mado Adhi and Waye Godha, land fragmentation caused by farming and range enclosures influenced access to forage resources. The cattle population in Mado Adhi showed stability for 13 out of 21 years but declined in 3 out of 21 years. The highest declines compared to pre-drought were in 1991/92 (–800%), 2006 (–150%), and 2011 (–98%).

The households in Waye Godha experienced lower numbers of cattle mortalities, except for the 1991/92 and 2011 droughts. Like other sites in the Waye Godha, total cattle populations showed less variability. There were no herd declines in 9 out of the 21 years. The highest declines occurred during the 1991/92 drought (–96%) and 2011 (–300%). There was a partial herd recovery in 2008.

The second population pattern-marked by the dynamics of reproductive females and the calves is represented by the Sololo-Ramata (arid/semi-arid agroclimatic zone), the Mukutano and Damballa Fachana (very arid agroclimatic zone) sites. Total cattle populations across the three areas varied over time (Fig. 7.5A, B and C). To begin with, the reproductive females fared better than those of other classes across the three sites. Calves, immatures experienced moderate population crashes, and mature males relative to their pre-drought population levels. All categories of cattle experienced recovery before plummeting to the lowest densities after the 1991/92 drought. There were moderate recovery levels in 1996/97 before declining during the 2000/01 drought (except the Damballa Fachana site). Cattle populations in the Sololo-Ramata and Damballa

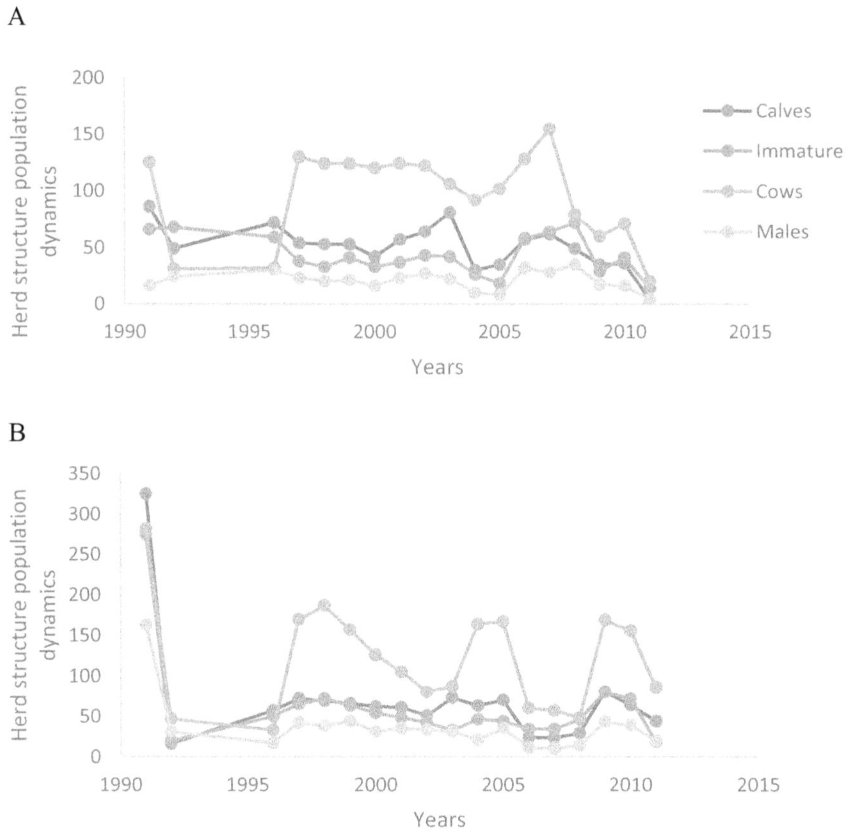

Fig. 7.4 Herd structure population dynamics in (A) Mado Adhi and (B) Waye Godha

Fachana sites disclosed moderate recovery during the 2008/10 period when the population of immatures exceeded those of calves. By 2011, the population numbers in all cattle categories had collapsed.

In the Sololo-Ramata site, the total cattle population declined by 400 percent during the 1983/84 drought and 600 percent during the drought of 2011. The cattle population collapsed in 6 out of 29 years and showed evidence of recovery in 8 out of 29 years.

At the Mukutano site, declines occurred during the droughts of 1983/84 (–150%), 1991/92 (–700%), 1997 (–120%), 2000/01 (–200%),

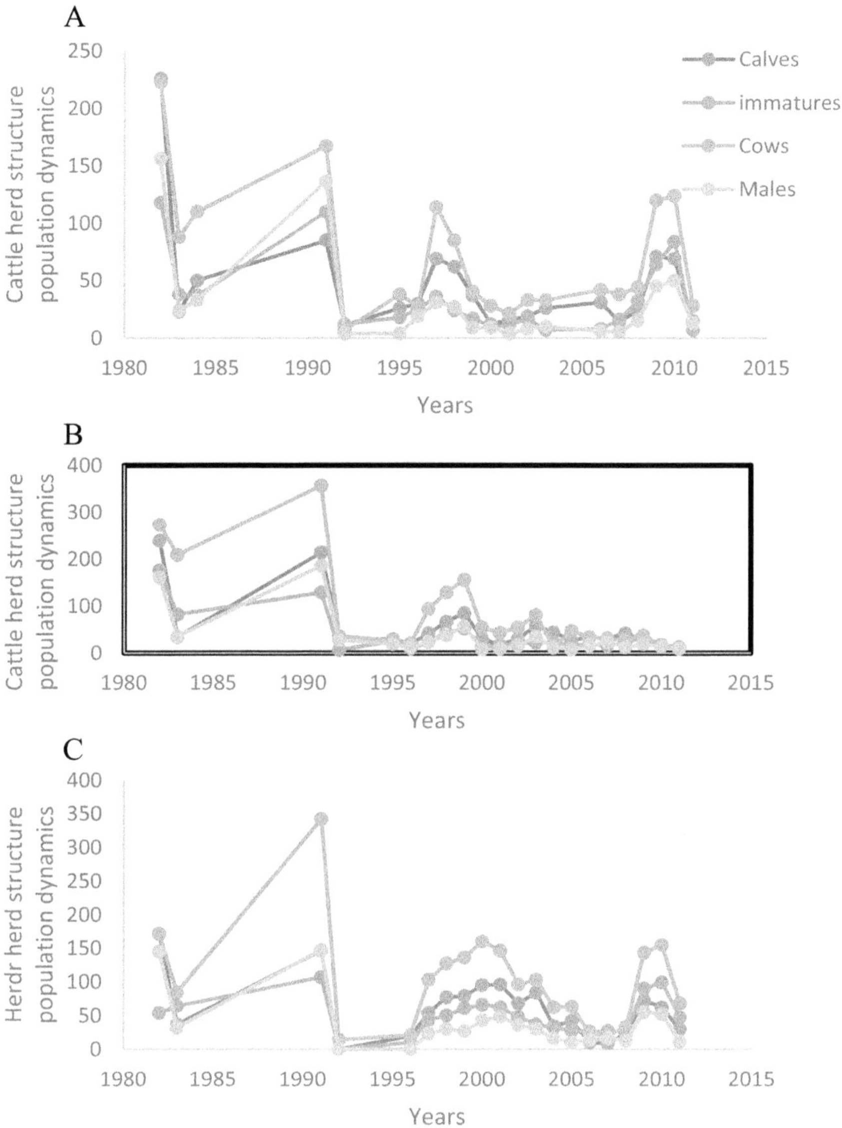

Fig. 7.5 Herd structure population dynamics in (A) Ramata, (B) Mukutano, and (C) Damaballa Fachana

and 2011 (–200%). Total cattle populations in this locality declined in 7 out of the 29 years. For 8 out of 29 years, there were no changes in entire cattle populations, while in 5 out of 29 years, cattle populations showed evidence of partial recovery.

Conversely, in the Damballa Fachana site, the most significant reductions in total herds occurred during 1983/84 (–150%) and 2011 (–150%) droughts. Declines in entire cattle populations occurred in 1991/92 (–150%), 2001/02 (–100%), and 2006 (–120%). Total herd numbers remained unchanged in 5 out of the 29 years, while in 7 years, there were partial recoveries. The third population pattern is represented by the sites of Anona, Uran, and Golole (Fig. 7.6A, B and C). The first two sites are in the semi-arid agroclimatic zone. The last site is in the arid agroclimatic zone.

The similarities among the sites are that farming and private range enclosures fragmented the grazing landscapes. Cattle from these locations traveled long distances to reach pasture during the droughts of both 1983/84 and 1991/92, causing high mortalities. Apart from the drought years, the reproductive females showed higher survival numbers across all three sites—except 1992–1994. The mature males had lower mortalities and persisted at lower densities. In the recovery period of 2006–2010, the immature cattle class showed comparable recovery with the calves.

In the Anona site, total cattle populations did not decline in 12 out of the 21 years. The highest declines occurred during the 1991/92 drought (–900%). Other reductions occurred between 2000 (–200%) and 2011 (–300%). In the Uran site, the total cattle population showed stability in 16 out of the 21 years. Cattle demography in the Anona site collapsed in 1991/92 (–1000%), 2000–2001 (–200%), and 2011 (–400%). Partial recovery occurred in 1996 and 2006. By comparison, in the Golole site, the cattle population collapsed during the 1991/92 drought (declined by 3500%), followed by a partial recovery in 2008. In 15 out of the 21 years, the cattle population remained stable. In all the reported sites, the extent of herd mortalities influenced patterns of recoveries.

Herd Recoveries

Between 1982 and 1990, herd recovery had achieved a maximum of 614 percent of the pre-1982 drought population. At this growth rate of 12.4 percent per year, recovery was achieved in 16 years after the 1983/84

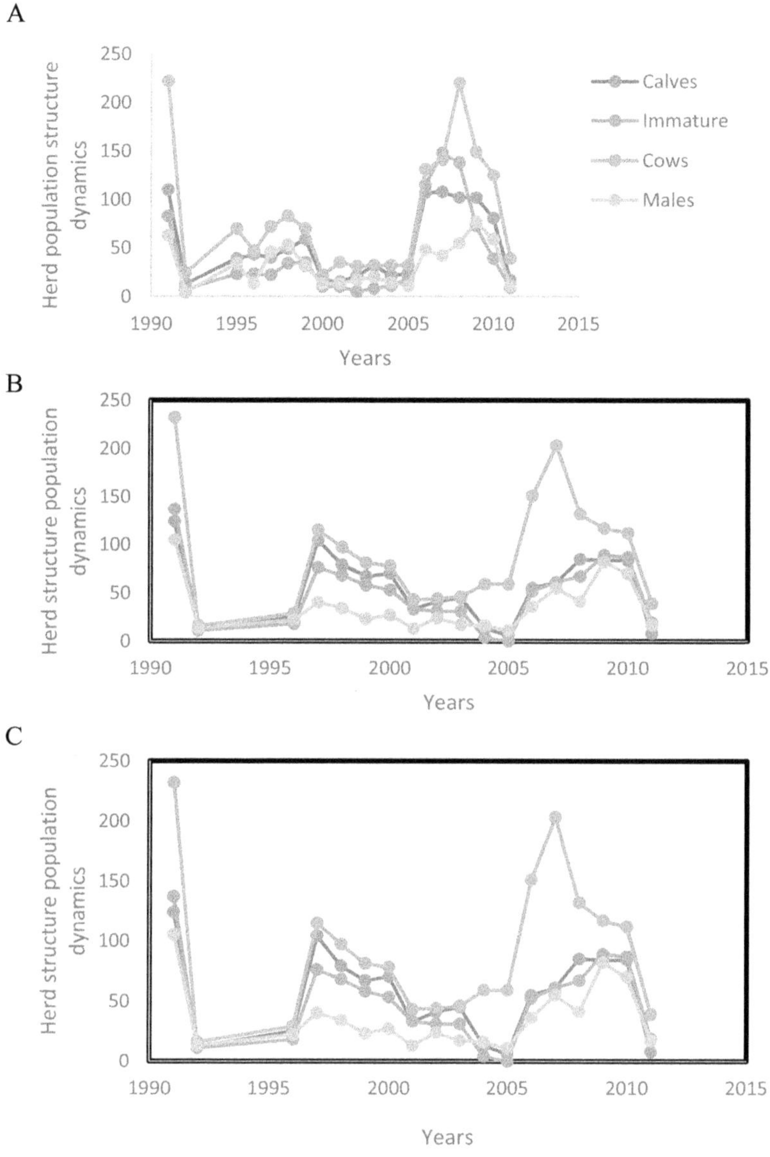

Fig. 7.6 Herd structure population dynamics in (A) Anona, (B) Uran, and (C) Golole

droughts. Relative recovery rates after the 1983/84 drought varied over time, from a maximum of 10.3 percent per year (1984–1985) to 1.1 percent per year (1989–1990), with an average herd growth rate of 3.5 percent per year. During the 1992–1996 recovery phase, the relative recovery rate was lower, varying from 4.7 percent per year (1992–1993) to 1.9 percent per year (1995–1996), with an average herd growth rate of 2.8 percent per year. The mean relative herd growth rates during the first recovery phase are comparable with the theoretical herd growth rate, while those of the second recovery phase are much lower. Cattle herd growth rates during the recovery periods are like those reported for the Sahelian zone of Africa.[50] Thus, different responses by regional cattle populations to various droughts and subsequent recovery periods can be attributed mainly to the seven recovery years (1984–1990) that saw above-average rainfall, resulting in increased grazing resources and calving rates. In a southern Turkana study, McCabe[51] reports no recoveries by the first year; among one herd, full recovery was achieved three years after the drought, and in another herd, recovery occurred five years after the drought. Conversely, the mean relative growth rate of the regional cattle population after the 1991/92 drought was 2.8 percent per year, which, after five years (by 1996), achieved a recovery rate of only 32 percent.[52] The recovery process was broken by the third, fourth, and fifth cattle population crashes in 2001–2002, 2004–2005, and 2011 (Fig. 7.7).

Several factors influenced herd recovery following stress periods. First, herd recovery was affected by the calving rate, which averaged 63 percent before the droughts, while it was less than 5 percent during the droughts.[53] These calving rates are comparable with those reported elsewhere.[54] Second, the 1983/84 drought caused fewer calves and breeding females mortality compared to the 1991/92 drought. Third, the 1991/92 drought was followed by five years (1996) of above-average rainfall, yet

[50] Wagenaar, K. T. *Productivity of transhumant Fulani cattle in the inner Niger delta of Mali*. Vol. 13. ILRI (aka ILCA and ILRAD), 1986.

[51] McCabe, J. T., 'Drought and recovery: Livestock dynamics among the Ngisonyoka Turkana of Kenya,' *Human Ecology* 15 (4) (1987): 367–388.

[52] Oba, G. 'The effect of multiple droughts on cattle.'

[53] Unpublished data.

[54] McCabe, J. Terrence. 'Drought and recovery: livestock dynamics among the Ngisonyoka Turkana.'

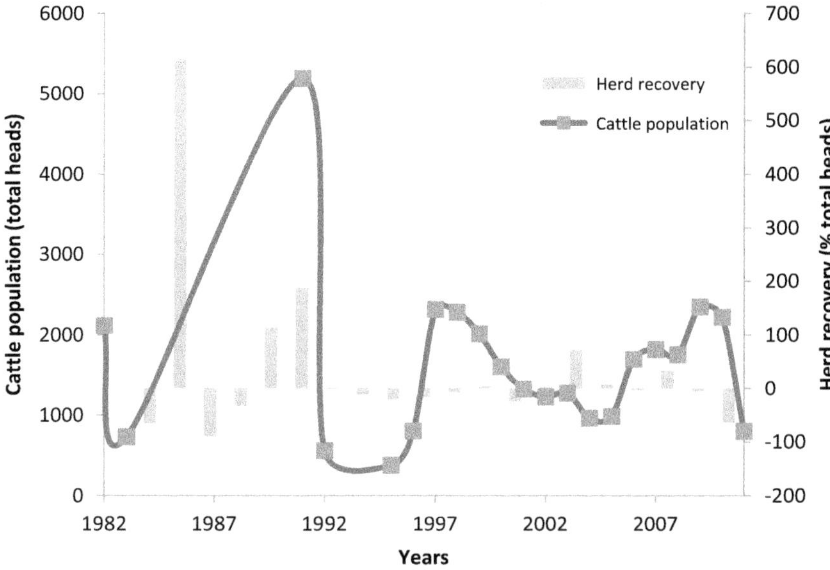

Fig. 7.7 Comparison of total cattle population and herd recoveries

herd growth needed to reflect the extent of available forage. This incomplete recovery was attributed to low numbers of surviving breeding females. Herd mortalities had implications for shifts in household wealth.

Tracking Household Wealth

Four critical considerations exist in the life history analysis of household wealth dynamics. First, cattle are susceptible to climate variability as a form of reproductive capital. Second, household wealth varies considerably across time. Third, household wealth tends to collapse during periods of drought, with incomplete recovery in most cases. Fourth, wealth is transitory, shifting from one rank to another depending on levels of herd mortality. Household wealth is tracked in terms of cattle per capita, changes in household wealth associated with cattle mortalities, and household wealth life histories. We discuss each consideration in turn.

Table 7.2 Obbu Borana households' cattle per capita shows trends before and after the 1983/84 and 1991/92 droughts

Sites	*Pre-1983*	*Post-1984*	*Pre-1991*	*Post-1992*	*1996*
Anona	5.7 ± 2.8	0.9 ± 1.5	4.4 ± 3.4	0.5 ± 0.5	1.4 ± 1.6
Damaballa Fachana	4.9 ± 3.6	1.7 ± 1.3	6.0 ± 6.0	0.1 ± 0.2	0.8 ± 0.8
Gollole	7.9 ± 6.3	0.9 ± 0.8	7.0 ± 7.7	0.2 ± 0.2	0.8 ± 0.5
Sololo-Mukutano	6.0 ± 5.9	2.1 ± 1.5	10.8 ± 6.5	0.9 ± 0.9	1.4 ± 1.7
Sololo-Ramata	5.2 ± 5.3	1.1 ± 2.5	5.3 ± 6.6	0.3 ± 0.3	0.9 ± 0.8
Madho Adhi	n/a	n/a	10.0 ± 7.7	0.8 ± 0.6	1.3 ± 1.5
Waye Godha	n/a	n/a	1.9 ± 2.3	1.3 ± 0.8	2.0 ± 1.8
Uran	6.1 ± 5.9	2.2 ± 3.0	4.9 ± 6.6	0.4 ± 0.3	0.9 ± 0.7

n/a, Data was not collected

Cattle Per Capita

Droughts and epidemics spurred declines in the number of cattle per capita[55] in many sites studied (Table 7.2). On average, the mean number of cattle per capita before the 1983/84 drought (5.98 ± 4.73) compared to post-drought (1.49 ± 1.76) declined by 75 percent. Before and after the 1991 drought, household cattle holdings elsewhere disclosed comparable wealth change patterns.[56] In a household survey in southern Ethiopia, Mengistu[57] shows that cattle holdings declined by about 74 percent. Many of those households already had numbers of cattle per capita far below the recommended subsistence level. Similar declines in cattle holdings have been reported for other pastoral communities in northern Kenya.[58] By 1996, five years into recovery, the number of cattle per capita in the Obbu Borana households had not recovered (Table 7.2) and was insufficient to

[55] Desta, S. and Coppock, D. L., 'Pastoralism under pressure: tracking systems change in southern Ethiopia,' *Human Ecology* 32 (4) (2004): 465–485.

[56] Cossins, N.J. and Upton, M. 'The Borana pastoral system of southern Ethiopia,' (*Agricultural systems* 25 (1987), 199–218).

[57] Mengistu, A. '*The Borana and the 1991–92 drought: A rangeland and livestock resource study*,' Institute for sustainable Development, French Catholic Committee against Hunger and for Development, Addis Ababa, 1997.

[58] For the East African herders see also, McCabe, J. T., 'Drought and recovery: livestock dynamics among the Ngisonyoka Turkana of Kenya' (Human Ecology 15 (1987): 371–385), Merryman, J. T., 'The economic impact of war and drought on the Kenyan Somali,' (*Research in Economic Anthropology* 8 (1987): 249–275), Homewood, K. and Lewis, J. 1987 'Impact of drought on pastoral livestock in Baringo, Kenya,' *Journal of Applied Ecology* 24 (1987): 615–631); Lind, J. and Barrero, L. R., *Into the fold: What pastoral responses to crisis tell us about the future of pastoralism in the Horn*, Working Paper 091, Future Agricultures, www.future-agricultures.org.

support the human population. The findings suggest that the recovery of cattle per capita was significantly impaired.

Changes in Wealth Ranks

Differences in herd sizes resulted in different wealth ranks. Ordinarily, wealthy households—despite comprising less than 10 percent of the total number of homes—had the most significant total number of head of cattle. Household cattle wealth ranks varied over time, depending on the stresses encountered.[59]

Traditionally, the Obbu Borana estimated that the poor and very poor categories should comprise less than 20 percent of the population. We pose the question: did droughts increase wealth inequalities among households? Let us consider the evidence. Wealthy families disappeared after the 1983/84 and 1991/92 droughts, with household wealth dominated by the poor and very poor. In our analysis, the proportion of the very poor and poor increased from 54 percent after the 1982/83 drought to 97 percent after the 1991/92 drought.

During post-drought periods, the wealthy categories disappeared or were insignificant. The shifts were more dramatic during and after the droughts of 1991/92, 2000/01, 2004/05, and 2011 (Fig. 7.8). We found no evidence that individual households remained at their pre-drought wealth ranks. Instead, individual households shifted to lower wealth levels. Indeed, 24 percent of the households were stockless and at risk of dropping out of pastoralism.

Our finding contradicts the evidence presented in the literature for other East African pastoralists.[60] Most of these studies describe households retaining their status instead of shifts in wealth ranks. According to that viewpoint, rather than playing an equalizing role in household wealth, droughts created more significant disparities[61] in that the rich households

[59] Suter, J. W., 'Cattle and inequality: Herd size differences of pastoral production among the Fulani of Northeastern Senegal,' *Africa* 57 (2) (1987): 196–218.

[60] E.g., Fratkin, E. and Roth, E. A., 'Drought and economic differentiation among Ariaal pastoralists'; Nganga, S. K., Bulte, E. H., Giller, K., Rufino, M.C. and Herero, M., *Exploring use of livestock wealth and social capital by pastoral and agropastoral households in ASALs as insurance against climate change and variability risks: A case study of Samburu District in Kenya*, Report No. 309-2016-5185, 2013.

[61] Fratkin, E. and Roth, E. A., 'Drought and economic differentiation among Ariaal pastoralists of Kenya,' *Human Ecology* 18 (4) (1990): 385–402.

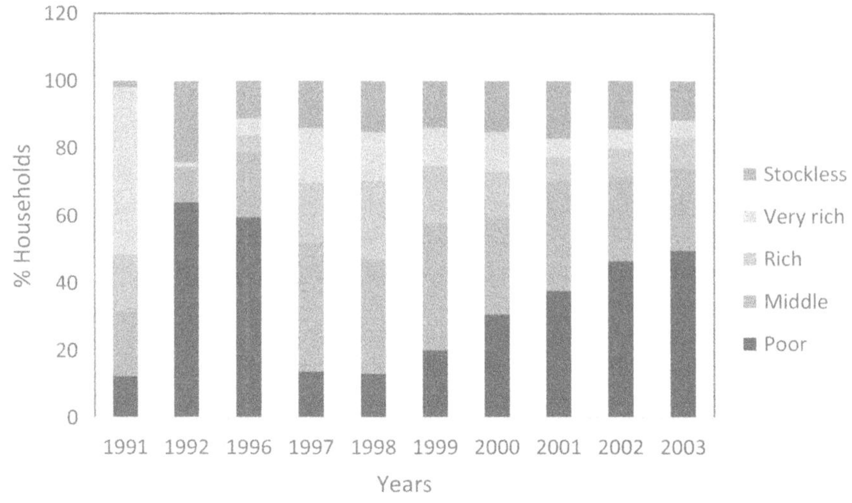

Fig. 7.8 Percent changes in household wealth ranks between 1982 and 2011

accumulated stock while the poor became poorer.[62] Consequently, larger herds served as security against drought, while those with few stocks were unlikely to survive in pastoralism.[63] Among the Arial pastoralists, for example, the rich and medium wealth households remained unchanged, while the poor became poorer.[64] Desta and Coppock[65] report that among the Borana of southern Ethiopia, a net dropout of 25 percent among the wealthy and middle-class, while 60 percent of poor households dropped out among Borana households. Like our findings, Somali pastoralists' poor and very poor families in eastern Ethiopia had increased after

[62] Lybbert, Travis J., Christopher B. Barrett, Solomon Desta, and D. Layne Coppock. 'Stochastic wealth dynamics and risk management among a poor population,' *The Economic Journal* 114, no. 498 (2004): 750–777.

[63] Hazell, Peter BR, and Ulrich Hess. 'Drought insurance for agricultural development and food security in dryland areas' *Food Security* 2.4 (2010): 395–405.

[64] Fratkin, E. and Roth, E. A., 'Drought and economic differentiation among Ariaal pastoralists of Kenya.'

[65] Desta, Solomon, and Coppock, D. L., "Cattle population dynamics in the southern Ethiopian rangelands, 1980–97." *Rangeland Ecology & Management/Journal of Range Management Archives* 55.5 (2002): 439–451.

droughts, while the medium wealth rank had disappeared.[66] We will demonstrate this using six cases of household wealth life histories among the Obbu Borana households.

Household Wealth Life Histories

We monitored the wealth life histories of six households belonging to Tucha Malicha, Dida Qampicha, Jilo Liban, Shama Alio, Roba Galgallo, and Halake Guyo between 1991 and 2003. Wealth changes were assessed by years in terms of total cattle holdings (in TLUs) per household. The data show that no family remained in a single rank over the ten years. Instead, following losses of stock during droughts, the same home shifted between the rich and the poor rank. This supports our hypothesis that cattle wealth among pastoral households is transitory, and a one-time ranking of the family might not produce accurate information. We will use the six case studies to demonstrate changes in wealth from one year to another.

We begin with the household of Tucha Malicha (Fig. 7.9A). Before the 1991/92 drought, this household was considered by neighbors to be wealthy. However, tracking the cattle-based wealth of this family shows dramatic changes between 1991 and 2003. Despite the partial recovery in 1998, the household had transitioned through various wealth ranks to poor and very poor by 2003, with their cattle holdings falling slightly below the minimum requirement (i.e., 4.5 TLU per person). In the traditional system of social security networks, the wealth of this household had shifted from a potential giver to a receiver of assistance (see Chap. 8).

Our next case is the household of Dida Qampicha (Fig. 7.9B). This household also experienced highly dynamic shifts in wealth ranks over a decade. Before the 1991/92 drought, they were classified as a medium—collapsing to very poor after the drought. Following recovery (1996), the wealth rank of Dida Qampicha had improved before declining to very poor in 1998 and, after slight improvement, falling to the very poor once

[66] Kassahun, A., Snyman, H. A. and Smit, G. N., 'Impact of rangeland degradation on the pastoral production systems, livelihoods and perceptions of the Somali pastoralist in Eastern Ethiopia,' *Journal of Arid Environments* 72 (2008): 1265–1281; Tache, Boku, and Gufu Oba. 'Is poverty driving Borana herders in southern Ethiopia to crop cultivation?' *Human ecology* 38.5 (2010): 639–649.

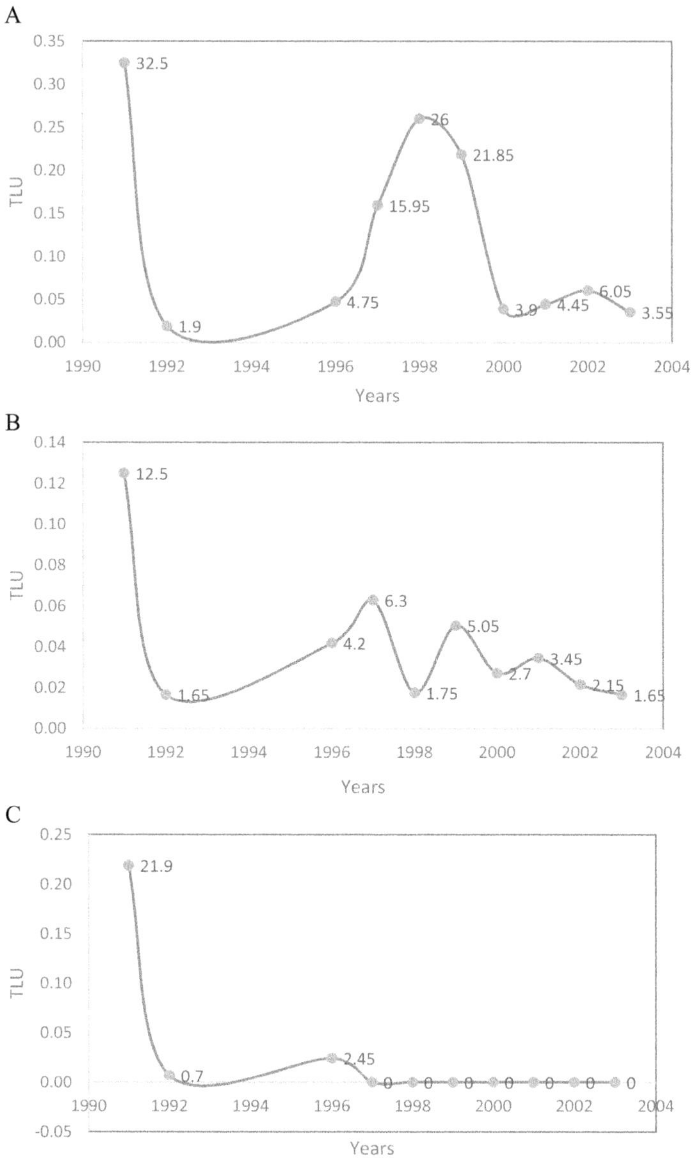

Fig. 7.9 Shifts in wealth (TLUs) from 1991 to 2003 belonging to the household of (A) Tucha Malicha, (B) Dida Qampicha, and (C) Jilo Liban

again. This household also shifted from being a potential giver to a receiver of stock (Chap. 8).

Our next case is the household of Jilo Liban, which shifted from a very rich to a stockless class after the 1991/92 drought. From 1997 to 2003, the family remained stockless and, for all practical purposes, had dropped out of pastoralism. Likewise, the household of Shama Alio (not shown) presents an interesting situation that might explain why stockless families keep hoping to return to pastoralism. Shama Alio was considered wealthy before the drought of 1991/92, after which the household was ranked very poor and stockless before recovering to the poor rank. Despite the family remaining below the pre-drought wealth ranking, their wealth trend is upward.

Similarly, the household of Roba Galgallo was ranked rich before the drought of 1991/92, soon declining to very poor. During the post-drought period, its wealth recovered, except during 2000–2001 when it became the poor category, before recovering to the medium rank by 2003. Finally, the household of Halake Guyo demonstrates the volatility of cattle-based household wealth. Between 1991 and 2003, the wealth ranks of this household shifted through decline and recovery. Before the drought, the household was ranked medium rich before falling into the stockless class. After that, the family gradually recovered, first attaining the very poor and reaching the wealthy category by the recovery period in 1997. Another decline followed, sliding downwards to the rich and medium wealth classes by 1998. By 1999, this household was poor, afterward showing a slight recovery before plummeting into poverty following the drought of 2002/03.

We can now sum up the findings from the six cases. Following the decadal droughts, all these Obbu households shifted from higher to lower wealth ranks. In particular, the drought of 1991/92 was a decadal marker, collapsing cattle-based wealth. Shifts in wealth ranks had social implications that individuals attempted to exploit, for example, through livestock transfer systems[67] (see Chap. 8). We now describe how resilience and collapse of cattle demographics and wealth dynamics might be explained.

[67] Moritz, M., 'Livestock transfers, risk management, and human careers in a West African pastoral system,' *Human Ecology* 41 (2013): 205–219.

Explaining the Resilience of Cattle Demography and Household Wealth

We identified five proxies to operationalize the concept of resilience related to cattle demography: herd structure, herd mortality, herd recovery, variable rainfall, and annual changes (+/–%) in total herd numbers as percentage differences between pre- and post-drought periods. The structure of a pre-drought herd population reflects herd growth. This period is associated with an accumulation of resources, favorable rainfall that promotes herd reproductive capacity, and recovery of household wealth.[68] Drought-related mortality triggers changes in the population of cattle age/sex classes. How rapidly recovery occurs depends on the severity of the drought and might take between 5 and 16 years. If drought frequencies are high, it may take longer to return to the regeneration and conservation phases—regarding resilience to cattle population changes. Regarding changes in household wealth, the proxy indicators declined following droughts, meaning that the wealth ranks of households shifted correspondingly downwards. In most cases, wealth recovery remained incomplete. The effects of droughts on household wealth were irrespective of pre-drought household wealth ranks. Most conspicuous was a 24 percent increase in the stockless population. The Obbu households are becoming poorer over time (Chap. 8).

[68] Galvin, K. A., Thornton, P. K., Boone, R. B. and Sunderland, J., 'Climate variability and impacts on east African livestock herders: The Maasai of Ngorongoro conservation area, Tanzania,' *African Journal of Range and Forage Science* 21 (3) (2004): 183–189.

PART III

Collapse and Transformation of Social Capital Networks

CHAPTER 8

Resilience of Social Capital Networks: Collapse and Transformation, 1991–2012

A significant attribute of resilience in pastoralism is the survival of social-political institutions for managing social capital networks. In their various forms, social capital networks provide social insurance against disasters.[1] They link individuals, groups of people, and neighborhoods who all share a common survival goal.[2] In such relationships, social capital functions at two levels: it forms social structures in terms of the organization from household to neighborhood levels (horizontal structures), while at the regional scale, it functions under hierarchical leadership organizations (vertical structures) to manage disasters. The criteria for herd sharing are household wealth.

Given that wealth among pastoralists can be highly fluid (Chap. 7), the obligations of individuals, family, and clan members, in general, are of paramount importance in working toward 'repairing' economic damage suffered by unfortunate households. The redistribution of stock functions through clan organizational structures and individual relationships that evaluate and distribute stock from the wealthy (i.e., the givers) to the very

[1] Bottrell, 'Dealing with disadvantage: resilience and social capital of young people's networks,' *Youth and Society* 40 (4) (2009): 476–501.

[2] Pferfferbaum, B., Van Horn, R. L. and Pferfferbaum, R. L, 'A conceptual framework to enhance community resilience using social capital,' *Clin Social J.* 45 (2017), 102–110.

G. Oba, *Pastoralist Resilience to Environmental Collapse in East Africa since 1500*, https://doi.org/10.1007/978-3-031-48291-5_8

poor (i.e., the receivers).[3] Consequently, where indigenous social capital networks provide insurance against disasters,[4] a breakdown of social capital networks occurs when there is insufficient stock for redistribution.[5] When this happens, the safety net fails to support those at risk of falling out of pastoralism.[6]

Accordingly, the ethnographic literature must explore why social capital institutions and networks fail systematically and over time.[7] In most cases, social capital analyses have been selective and short term.[8] This lack of long-term research fails to forecast why customary social capital activities fail and lose resilience. Additionally, following the collapse of traditional social capital networks, its transformation should be monitored more. Individual systems are rarely monitored over periods of disasters and recoveries. Further, the voices of principal actors in social capital network

[3] The system described in this chapter is different from the stock friend or bond friend relationships described for the Samburu by Spencer, Paul, *The pastoral continuum: The marginalization of tradition in East Africa* (Oxford University Press, 1998); For the Turkana, see, Storås F. (1991) "Cattle complex a begging complex. Livestock transaction and the construction of Turkana society" in Grunhaug, R., Haaland, G. and Henriksen, G. (eds.) *The ecology of choice and symbol Essays in Honour of Fredrick Barth* (Alma Mater Forlag, Norway, 1991); for the Ariaal, see Fratkin, E., *Ariaal pastoralists of Kenya: surviving drought and development in Africa's arid lands* (Ally and Bacon, Boston USA, 1998), for the Rendille and the Gabbra, see Schlee, G., *Identities on the move: clanship and pastoralism in Northern Kenya* (Manchester University Press, 1989).

[4] O'Leary, M. F., *The economics of pastoralism in Northern Kenya: The Rendille and the Gabra*. IPAL Technical Report No. F-3. UNESCO-Nairobi, 1985; Holden, S. J., Coppock, D. L., and Mulugeta Assefa, 'Pastoral dairy marketing and household wealth interactions and their implications for calves and humans in Ethiopia' (*Human Ecology* 19 (1991): 35–59; Oba, G. 'Kenya Borana: sharing and surviving, responses of impoverished pastoralists to food insecurity,' *Extension Bull. Of the University of Reading* 4 (1994), 17–22; Beyene, F., 'Interclan cooperation in a risky pastoral ecology: some lessons from Eastern Ethiopia,' *Human Ecology* 38 (1991): 555–565.

[5] My definition of social capital follows that of Rostila, M., 'The facets of social capital,' *Journal of the Theory of Social Behaviour* 41 (2010): 308–326.

[6] See Bollig, M. 'Moral economy and self-interest: kinship, friendship, and exchange among the Pokot (N.W. Kenya),' In Schweizer, T. and White, D.R. (eds.) *Kinship. Networks, and exchange* (Cambridge University Press, 1998).

[7] Campbell, D.J., Response to drought among farmers and herders in southern Kajiado District Kenya: A comparison of 1972–1976, *Human Ecology* 27 (3) (1999): 377–415.

[8] Santos, P. and Barrett, C. *Herd dynamics, social networks, and informal transfers among southern Ethiopian pastoralists*, Research Brief 05–07, 2005-PARIMA, Global Livestock CRSP.

decisions should be included in existing analyses.[9] We are particularly interested in links between wealth and the functional performance of traditional social networks operating at different scales. We aim to understand how the Obbu Borana responded to the failures of traditional social capital networks. Even more importantly, we are interested in how and why social interactions continue within national and across international borders (referred to as 'social capital network zones'). The two regions share historical, cultural, and economic relations. The Ethiopian side of the frontier is the traditional home base of the *gada*—the Oromo democratic institution responsible for administering clan social capital networks.[10] In all transactions, political circumstances and the economic situation along the frontier served as assets to the smooth functioning of social capital networks.

Therefore, these relationships should be investigated considering local and regional political factors that affect accessibility to social capital networks across the frontier. We are interested in understanding the functioning of social capital networks and their transformation in response to multiple droughts. We evaluated the performance of the following types of networks among Obbu Borana households: clan social capital networks (called *buusa gonoofa*), individual networks of sharing milk animals (called *daabare*), and the newly developed broad-based neighborhood social capital networks (called *baaliti*). We investigated specific methods used to organize fundraising in the neighborhood social capital networks. We considered how fluctuations in the number of participants affect the fundraising activities in the neighborhood social capital networks. We provide examples of how social capital networks were building blocks for transformation and resilience in pastoralism.

This chapter has four objectives: (1) understanding the context of using clan social capital networks. Here, we will review the criteria used for wealth distribution, evaluate the capacity of individual social capital networks, and understand their transformation. (2) We will then evaluate the performances of *buusa gonoofa* and individual networks of sharing milk

[9] Exception is Homewood, Katherine, Kristjanson, Patti and Trench, Pippa (eds). *Staying Maasai? Livelihoods, conservation, and development in East African rangelands* (Vol. 5. Springer Science & Business Media, 2009).

[10] Bassi, Marco. "The System of Cattle Redistribution among the Boraana Obbuu and its Implications for Development Planning.' In Baxter P.T. W. and Hogg, R. (eds). *Property, poverty, and people: changing rights in property and problems of pastoral development* (Department of Social Anthropology, University of Manchester, 990).

animals. (3) We will appraise the effectiveness of neighborhood social capital networks in terms of their performance using a neighborhood case study. (4) Finally, we will discuss the possibilities for building resilience in neighborhood social capital networks.

The Context of Using Social Capital Networks

Social capital networks are resource-sharing tools among kin, stock friends, sub-clans, neighborhoods, and individual networks. In the kinship system,[11] where mutual assistance links genealogically related individuals, the redistribution of resources is voluntary.[12] It describes the benefits individuals may accrue through positive social interactions.[13] It has elements of trust and reciprocity, which Ostrom and Walker[14] suggest is part of a sequence of mutual relationships displayed by groups sharing resources. Social capital networks, therefore, recognize the multi-nature of social roles[15] that function under different social rules, as defined by the criteria used for the poor to gain access to the wealth of well-off members. This would also include shifting resource sharing from a neighborhood to a regional level, where individuals exploit resources in the broader community.[16] We now seek a basic understanding of the ethnography of traditional Borana social capital networks.

Social Criteria for Wealth Distribution

The Borana pastoral economy measures wealth regarding cattle herd sizes (Chap. 7). The wealthy would have more than 50 heads of cattle. Those

[11] Bollig, Michael. *Moral economy and self-interest: Kinship, friendship, and exchange among the Pokot (NW Kenya)*. Cambridge University Press, 1998.

[12] Cleaver, Frances. "The inequality of social capital and the reproduction of chronic poverty." *World development* 33.6 (2005): 893–906.

[13] Lesser, E.L., 'Leveraging social capital in organizations,' In Lesser, E.L, and Wobarn, MA (eds.) *Knowledge and social capital: Foundations and applications* (Butterworth-Heinmann, 2000).

[14] Ostrom, Elinor, and James Walker, eds. *Trust and reciprocity: Interdisciplinary lessons for experimental research* (Russell Sage Foundation, 2003).

[15] Obstfeld, D. 'Social networks, the Tertius lungens, orientation, and involvement in innovation,' *Administrative Science Quarterly* 2005 https://doi.org/10.2189/asqu.2005.50.1.100 https://asq.sagepub.com/content/50/100.

[16] Lin, Nan. 'A network theory of social capital,' *The Handbook of Social Capital* 50.1 (2008).

in the medium category (*uufira buula*) have 25–50 cattle, and individuals categorized as 'weak-hands' (*harka qaaleesa*) owning 10–20 head are the next in line for stock redistribution. Those who own less than ten head of cattle are classified as poor (*deega*), the stockless (*qollee*) own no herd, and the perpetual poor category (*hiiyeesa*) are those who own less than two head and would have a waiver from clan contribution.[17] Those in the *qollee* (stockless) category are usually the primary beneficiaries of clan stock redistribution.

The restocking process proceeds in three stages. The initial stage of redistribution—akin to 'first aid' (*qullamo*)[18]—is the responsibility of the people of close kinship who distribute milk to the children and older people. In the second stage, individual sub-clans hold separate annual assemblies to evaluate the extent of poverty caused by disasters and levels of disposable stock among members within each social capital zone (*gaadisa*). The third stage involves clan participation at the regional scale (i.e., including the host social capital zones on the Ethiopian side of the border).[19] Before the host social network commits itself to discussing restocking needy families from the home social network zone on the Kenyan frontier, they inquire whether the first and second stages of restocking have been performed. Once the recommendations have persuaded the host network, they will agree to discuss restocking the stockless in the sub-clans on the Kenyan side of the border. After the participating councilors have observed all protocols, the clan assemblies request those with sufficient herds to contribute milk cows to some and heifers and male animals to others.

The host determines livestock availability and declares the numbers to be contributed. Suppose the host capital network has insufficient stock. In that case, it recommends that representatives of individual sub-clans from the Kenyan frontier proceed to the following social capital network zone, still on the Ethiopian side of the international border.

At the assemblies, the Obbu representatives—from the home social capital network—need to explain the reasons for presenting their cases to the host social capital network. The economic situation of the home social

[17] See Tache, Boku, and Espen Sjaastad. '"Pastoralists" conceptions of poverty: An analysis of traditional and conventional indicators from Borana, Ethiopia,' *World Development* 38.8 (2010): 1168–1178.

[18] see Bassi, Marco. 'The System of Cattle Redistribution among the Boraana Obbuu and its Implications for Development Planning.' (1990): 32–37.

[19] Helland, Johan. *Five essays on the study of pastoralists and the development of pastoralism*, 20 (1980).

network is explained without deviating from decisions reached earlier. The host social capital networks usually resist shifting responsibility onto their members, especially if there is any hint that the home social capital network contributions are not commensurate with their levels of disposable stock. If such failure were due to individuals refusing to give according to clan decisions, then force (*jiimfensa*) would be contemplated to collect the chosen stock from the deviant member. Besides clan social capital networks, friendship relationships are used for individuals to build networks by exchanging milk animals.

Evaluating Individual Social Capital Networks

Unlike in the sub-clan social capital networks, where giving is obligatory, in individual stock-sharing networks, the sharing of milk animals is negotiated. The practice of loaning milk cows (*daabare*) to friends and relatives in need is insurance against milk shortages for young children. In this system, the obligation to give is not time-bound but is flexible, depending on social and economic conditions. In essence, individuals build their social security networks that include bond friendship networks and those among relatives. In bond friendship relations, individuals mutually establish their rights. Such social security networks operate optimally only if members continuously build new ties and repair old ones.

The participants in a *daabare* network are members of a patrilineal group, people of the same age set, or those related through marriage or networks of friends that cut across clan alliances.[20] The *daabare* contract is mutually binding but should be differentiated from obligations claimed as a right as in *buusa goonofa* clan social security networks. The stock transfer period may vary, depending on the food situation of the needy family. The rules are that ownership rights to the loaned animal are maintained, including all female progeny, by ear clipping and branding with the mark of the primary owner. Conversely, all male calves belong to the borrower.[21] If the animal dies, the borrower is not obligated to replace it, although

[20] For the Arrial and Rendille see, Fratkin, E. *'Ariaal pastoralists of Kenya: Surviving drought and development in Africa's arid lands'* Allyn and Bacon, Boston, USA, 1998. For the Rendille and Gabbra, also see, Schlee, Günther 1989, *Identities on the move: Clanship and pastoralism in Northern Kenya*. Manchester University Press.

[21] See Dahl, Gudrun. *Suffering grass: Subsistence and society of Waso Borana*. Diss. Stockholm University, 1979.

they are obliged to use loaned animals only for purposes authorized by the primary giver.

Following the collapse of pastoralism after the 1991/92 droughts, the clan and individual social capital networks also collapsed. The society was obliged to transform the system into what they called 'broad-based' neighborhood social capital networks (*baaliti*).

Transformation of Social Capital Networks

The destruction of herds caused by regional droughts resulted in declining household cattle wealth (see Chap. 7). Economic fortunes shifted, with former wealthy households becoming poorer or even stockless. Traditional indigenous social capital networks also failed, which motivated the establishment of a new system of broad-based neighborhood social capital networks.

The changes created a material shift from loaning cattle to making monetary contributions. In the new system, both the wealthy and the poor are actors. These neighborhood networks assist the most disadvantaged members by creating common reservoirs of resources.[22] The criteria for sharing had also changed. Poverty in stock is not the only factor—but others include assisting neighbors with school fees and paying hospital bills, among numerous other human needs. Each neighborhood might also be bonded through other social relations that bind everyone (see Chap. 9). Although the physical arrangements of the areas might create an impression of economic uniformity,[23] differences may exist[24] in terms of types of resources, populations of vulnerable people, and how they might respond to disaster management.[25] The mushrooming of registered self-help groups (e.g., women's groups, youth groups, etc.) and community-based organizations partnering with development projects has created various social capital networks. Help to members is through public

[22] Aldrich, D. P., *Fixing recovery: Social capital in post-crisis resilience.* Purdue University-Purdue-Pubs 5-2010, Department of political science.

[23] Aldrich, D.P., Meyer, M. A., 'Social capital and community resilience,' *American Behavioral Scientist* 59 (2) (2015): 254–269.

[24] Dynes, Russel, R., *Community social capital as the primary basis for resilience.* Preliminary Paper # 344, 2005.

[25] Pfefferbaum, Betty, Richard L. Van Horn, and Rose L. Pfefferbaum. 'A conceptual framework to enhance community resilience using social capital.' *Clinical Social Work Journal* 45.2 (2017): 102–110.

fundraising, in which all neighborhoods are linked through mutual arrangements. We will continue by examining the performance of clan and individual social capital networks and why they collapsed before returning to the broad-based neighborhood social capital networks.

Evaluation of the Performance of *Buusa Gonoofa* Social Capital Networks

Whereas according to tradition, individuals with 50 animals or more would be considered moderately prosperous in the past, currently, they would be classified as wealthy among the crowded poor. On average, poor households have stock holdings of 0.84 TLU per household, below the minimum of 4–5 TLU per capita needed for household subsistence.[26]

Restocking stockless households had specific targets corresponding to the basic needs of a pastoral family. Ideally, the goal was to establish a human-to-stock ratio 1:1.[27] In practice, this was only possible sometimes. The clan's purpose was to attempt to provide a basic entitlement of five animals—called *shaanani goosa*—'the fives of the clan.'[28] These included three milk cows, one heifer, and one young male animal, according to 45.5 percent of the respondents. At a minimum, each stockless family is entitled to one male animal, which is required to meet burial expenses, according to 73.3 percent of the informants.

The community considered other causes of poverty that did not qualify for restocking. These included mismanagement and over-consumption of the herd through 'irresponsible actions' such as gambling and drinking alcohol from the income from cattle sales, according to 72.4 percent of the informants, or poverty related to the loss of non-pastoral economic activities (13.8%). If individuals expended their stock, the clan might, in principle, refuse to act, but in practice, the dependents were always given consideration. However, such assistance would be subject to stringent conditions. Specific conditions would also apply to a person who had refused to participate in stock redistribution in the past (3.4%).

[26] Pratt, D. J. and Gwynne, M. D., *Rangeland Management and Ecology in East Africa* (Hodder and Stoughton, London, 1977).

[27] Interview with Abduba Boru Sora by Gufu Oba in July 1992 in Obbu Kenya.

[28] The implication is that a destitute person who lacks access to milk will bleed his cows to feed his or her children. This is currently symbolic, rather than describing a widespread practice.

We scrutinized the stages of the restocking process (described earlier) in the post-drought periods of 1996 and 2005. By 1996, about 60 percent of the clans failed to participate in the first stage of restocking. Among the reasons for low participation were widespread poverty, according to 98.8 percent of the informants, border closure with Ethiopia (4.5%) due to insecurity, and a lack of consultation between clan leaders in the home network zone (3.7%). Ninety percent of the clan councilors we interviewed emphasized system failures.

Narratives of Clan Councilors

Only 10 percent of the Obbu sub-clans successfully met the ideal three-stage restocking process during the 1996 post-drought period. The following responses were given by senior and junior councilors of the sub-clans that partially succeeded in restocking people experiencing poverty. Jaldesa Boru Elema,[29] the senior *jalaab* of Karayu Akhaaro, narrated his experience as follows:

> After the 1991/92 droughts, we had 36 *qoollee* [stockless households] in the *gaadisa* of Obbu. The remaining families were left with 1 to 4 cows each. Those with 3 or 4 cows were asked to donate one, but those with 1 to 2 were exempted... In seeking support for the poor, we visited different *gaadisa* [on the Ethiopian border]. Our sub-clan in Obbu was united with the *gaadisa* of Badha [the humid zone in Ethiopia] and Moyale. Two of us were sent to the *gaadisa* of Mega, Dubluq, Malbe, Yaabalo, and Erdar [all in Ethiopia] in that order... In the *gaadisa* of Mega, we met the *abba qae* [of the host social capital zone] and presented our case. The *abba qae* in Mega representing our clan was obligated by custom to call a clan assembly. After further consultations, we were told that our sub-clan in that social network zone was equally in dire economic stress. They were responsible for forwarding our case to the next *gaadisa* and explaining their inability to restock the poor [from the Obbu social capital network zone] ...The *abba qae* in Mega took us to the *gaadisa* of Yaabalo [which was the next host social network zone]. Here, we found the *hayu* of our clan. He enquired about the purpose of our visit and the clan's welfare in the *gaadisa* of Obbu...The *abba qae* of the *gaadisa* of Yaabalo asked his counterpart in Mega whether he was satisfied with our presentation. On that basis, the clan annual assembly agreed to discuss our case.

[29] Interview by Gufu Oba in August 1996, in Ramata, Kenya.

According to the source, four families in Obbu were restocked from the home social capital network zone during the second stage. At Mega (for the third stage), cases of 36 destitute households were discussed. Jaldesa Boru Elema continued with his narration:

> Our *hayu* accompanied us to the *gaadisa* of Malbe [in Ethiopia]. From this *gaadisa*, we received 18 additional cows and sold one.[30] All in all, we returned to Obbu with 21 animals. We called for another assembly [back in the home network zone], but this time to report on the stocks we received and to discuss how to redistribute them. The animals received were fewer than the number of stockless households.

The sub-clan had a deficit of 15 households that had met the criteria for restocking. Jaldesa Boru narrated their next move:

> We postponed the annual clan assembly for the coming short rainy season in 1994. When the short rains arrived, the political situation on the border [with Ethiopia] had worsened, and traveling across had become unsafe. We again postponed the meeting until the long rains [April–May] 1995. We sent information about the number of stockless needing to be restocked. Still, knowing that we could not move from one *gaadisa* to another, we asked the participants from the home social network zone to contribute cash instead.[31]

Here, one may infer how the society transformed the indigenous restocking tradition from stock gifts to cash payments. We use another example.

The sub-clan of Karayu Danqa experienced partial success in restocking the poor with the help of their Ethiopian counterparts. According to Qallicha Charfana Diido,[32] in the social network zone of Obbu, there were 73 destitute households of the clan, but including people with only 2 to 3 cows brought the total number of families that qualified for restocking to 80. The sub-clan had lost the capacity to supply people experiencing poverty during the first and second stages of the restocking process and, therefore, forwarded their cases to the assembly of the *gaadisa* of Goorille

[30] This animal—referred to as *saa qoop'ee* (the animal of the footwear)—is an entitlement to the messengers to cover their travel expenses.

[31] This was the beginning of the practice of neighborhood broad-based social capital networks.

[32] Interview with Qalicha Charfana Dido, senior abba Qae of Karayu Danqa 4 August 1996 in Madho Adhi, Kenya.

on the Ethiopian border. Members of the social network of Goorille were equally impoverished, with no possibility of restocking stockless households from the home social capital network zone. They, therefore, forwarded the cases to the social capital network zone of Dubuluq, which in turn referred them to the Goomole zone in Ethiopia at the ritual settlement of the *Qallu*. There, the Obbu delegates received 12 cows, with a promise of additional animals in the future. Their case had taken more than four months and concluded with limited success.

Similarly, the sub-clan of Karayu Tissele—despite contact with their counterparts on the Ethiopia side of the border—failed to restock the destitute. The senior councilor of this sub-clan—Philip Bule Gufu, a primary school teacher, narrated his sub-clan experience as follows:

> After the droughts of 1991/92, we organized an annual assembly in our *gaadisa* of Obbu, including Tuka, on the Ethiopian side of the border. After we completed the first stage of assisting the impoverished members, we sent the cases of 170 stockless households to the *gaadisa* in [southern] Ethiopia. One *jaalab* and an elder presented our case. Given that our sub-clan in [southern] Ethiopia had experienced dire economic problems, our delegates were referred to the ritual settlement of the *Qallu* in Goomole in Ethiopia. We received 23 animals (mostly adult females and heifers), which did not meet the needs of low-income people in our sub-clan. In distributing the few animals, we prioritized households with more dependents and older people. We planned to return to the *gaadisa* in Ethiopia during the recovery phase but could not do so because of insecurity.[33]

Other sub-clans did not fare any better. The Dambe-Noono alliance and Digaalu (Eemaji, Tiiti, Nurtu, and Uduumtu) were among the sub-clans that failed in the three stages of restocking. According to the councilor[34] of the Dambe-Noono clan alliance, the clan social capital in Obbu had been unable to function, partly due to the collapsed pastoral economy and partly due to the lack of access to the Ethiopian side.[35] Equally, according

[33] Interview with the late Philip Bulle Gufu Jillo, senior abba qae of Karayu Beere in Sololo-Ramata on 17 July 1996.

[34] Interview with Roba Hanti Sap'ansa about their experience, by Gufu Oba in Sololo-Ramata on 20 July 1996.

[35] Oba, G. '"Where the bulls fight, it is the grass that suffers:" impact of border administration on drought-coping strategies of the Obbu Booran during the 20th century.' *Journal of Oromo Studies* 7.1&2 (2000): 87–108.

to Abdub Boru Sora,[36] the clan social capital network (*buusa gonoofa*) of the Digalu alliance had failed for several decades. As opposed to during the times of their parents (when people experiencing poverty looked upon the sub-clans for restocking), in recent times, men—particularly younger men—are joining the labor migration movement to seek salaried jobs and send remittances to support their families. The failure of networks also occurred at household levels.

Performance of Networks by Individual Households in Different Networks

Between 1992 and 1996, 57.3 percent of households failed to participate in *buusa gonoofa*, while the remaining 42.7 percent attempted to do so. Only 27.6 percent of the total number of families reported having received stock gifts from the sub-clans. The extent of participation in *buusa gonoofa* and *dabaare* networks varied across the communities (Table 8.1). While the communities in Madho Adhi were highly successful in participating in *buusa gonoofa*, those in the peri-urban environments of Sololo-Ramata were the least successful. Being cattle people, the Obbu Borana failed to include other animal species in herd distribution; thus, they lacked flexibility in reconstituting the poor after droughts had killed cattle.

Table 8.1 Participation (%) in *buusa gonoofa* and *dabaare* social capital networks by Obbu Borana households in eight locations between 1992 and 1996

Locations	*Buusa goonofa (%)*		*Dabaare (%)*	
	Yes	*No*	*Yes*	*No*
Anona	53.6	46.4	17.9	82.1
Dambala Fachana	61.1	38.9	0	100
Golole	35.0	65.0	5.0	95.0
Madho Adhi	92.0	8.0	8.0	92.0
Mukutano	45.9	54.1	21.6	78.4
Sololo Ramata	10.52	89.4	14.0	86.0
Uran	42.9	57.1	35.7	64.3
Walda	31.6	68.4	42.1	58.0

[36] Interview with Abdub Boru Sora by Gufu Oba in Sololo-Ramata on 24 July 1996.

Table 8.2 Correlation matrix between the size of stock (STOC), family size (FAMSIZ), participation in *buusa gonoofa* (BGOFA), receiving from relatives (REREL), giving to relatives (GIVREL), and participating in *dabaare* (DABAARE) by the households in Obbu

Factors	*STOC*	*FAMSIZ*	*BGOFA*	*REREL*	*GIVREL*	*DABAARE*
FAMSIZ	*					
BGOFA	***	**				
REREL	ns	ns	***			
GIVREL	***	*	***	ns		
DABAARE	**	ns	*	*	***	

ns = not significant, * $p = 0.05$, ** $p < 0.005$, *** $p < 0.0001$

After the 1991/92 droughts, 44.9 percent of the households provided *daabare* milk cows to friends and relatives, while 55.1 percent failed to do so. Apart from two rural communities (Walda and Uran), the *dabaare* networks failed (Table 8.1). The impression one gets is that all the social capital networks practiced by the Obbu Borana needed to perform better. What would the reasons be? Understandably, stock availability was critical for families to share milk animals with friends and relatives (Table 8.2). For the Obbu households, self-restocking (replacing lost animals) through animal sales was common. After the 1991/92 droughts, 27.4 percent of the households sold small stock, 46.3 percent sold cattle, and 26.3 percent sold poultry. They used the money to pay for clan assistance instead of cattle, among other needs. During the recovery phase 1996, many households failed to replace lost stock, with only 7.9 percent reporting successes. Many families (92.1%) said cattle prices at the local markets were prohibitive.

The data on choices of livestock species bought and sold might suggest that the Obbu Borana had few alternatives. Whereas it was a tradition that mature cattle were replaced with small stock,[37] the lack of relationship between cattle sales and purchases of small ruminants probably indicates that cattle replacements failed for most surveyed households. The evidence points toward the failure of both *buusa gonoofa* and *dabaare* social capital networks. The society responded by creating a new and novel broad-based neighborhood social capital network (*baaliti*). This hybrid

[37] Coppock, D.L., *The Borana plateau of southern Ethiopia: Synthesis of pastoral research, development, and change, 1980–91*. Vol. 5. ILRI (ILCA and ILRAD), 1994).

system has wholly transformed contributions to those in need by monetizing the agropastoral economy.

Appraisal of the Effectiveness of Neighborhood Social Capital Networks

During the 2005 surveys, respondents suggested that the clan social capital networks had been transformed regarding social obligations, structural changes, and entitlements. A senior clan councilor[38] (referred to below as 'GBA') interviewed for this research made the following observations:

> GO: You mentioned that sharing of *buusa gonoofa* across the Ethiopian side of the border has been disrupted. What adaptive strategies did the Obbu Borana use to compensate for this loss?
>
> GBA: The loss of access to the Ethiopian side motivated the Obbu Borana to develop a new system where everyone contributes according to their economic capacity. The contribution has shifted from stock to monetary donations. Some may give Ksh 20, 50,100, or 200. No one is left out except in cases of extreme poverty.
>
> GO: Why was there a shift from stock to monetary gifts?
>
> GBA: There needed to be more cattle to go around. Currently, the economy is expressed in terms of cash transactions. For the few who had cattle, people avoided the waste of property by giving out a bull. They sold cattle or any other animals and gave out the money.
>
> GO: When *buusa gonoofa* was paid using cattle, the clan used force (*jiimfensa*) to punish rebellious members. What measures are being used under the new system if individuals refuse to contribute?
>
> GBA: Force is no longer applicable because it is against government policy. Those forced by the clan will go to the chief or the police to accuse the clan elders, saying that their property was taken by force [The alternative action against those who refused to pay is to deny them public support].

Huqa Morowa[39] underscored the fundamental weakness of the clan social capital network. The widespread impoverishment of the population had forced the Borana to consider changes in supporting the poor. Due to the diversified economy, individuals now have access to cash, which is given in two forms. The first method asks individuals to donate specific cash

[38] Interview with Gayo Boru Aga by Gufu Oba in July 2005.
[39] Interview with the late Huqa Morowa by Gufu Oba in July 1996.

amounts, symbolically called 'cattle.' The second method is to organize a public fundraising campaign.

The concept of 'wealth' has changed in the new social capital networks. Different criteria are used for redistributing resources in the changed social capital networks. The new system treats all participants as equals in accessing community-wide resources, whether wealthy or poor. This means that the definition of 'poverty' also appears to be changing. Historically, poverty was measured regarding lack of stock, especially cattle. Under the current circumstances, all income is defined as monetary gifts.

The clan members are participants in the neighborhood networks. Part of the support given is not included as a criterion in the clan assistance. These included service with school fees and providing support to the sick. We will give an example. An informant[40] stated: 'We participated at the beginning of the year in fundraising to pay school fees for a bright student who was admitted to a secondary school. Every clan member was asked to contribute KSh 50–100. We then requested public participation and together raised KSh 80,000.' This amount exceeds what would have been achieved by the traditional *buusa gonoofa* network support.[41] Further, we conducted a performance assessment of the new neighborhood social capital networks and noted dramatic changes.

Performance of the Neighborhood Social Capital System

During the 1996 survey, about 80 percent of households suggested that they respond to widespread poverty by establishing neighborhood social capital networks. During the 2005 surveys among junior and senior clan councilors, the shift from clan-based networks to broad-based *baalliti* social capital networks progressed through three stages. First, after the failure of restocking by clans, the community established clan alliances to pool resources. In the second stage, different clan alliances in the two opposite moieties (i.e., *Saabo* and *Goona*) jointly helped people experiencing poverty through public fundraising efforts. The third stage involved external development agents incorporated into the new social capital networks. Systematic methods are used to organize fundraising for those who request help.

[40] Interview with the late Huqa Morowa by Gufu Oba on 24 July 1996.

[41] Interview with Kula Jillo by Gufu Oba on 24 July 1996.

The networks operate through selected committees comprising a chairperson, secretary, treasurer, and members of both genders. The treasurer is chosen by the household that is requesting funds. The various neighborhoods select other members of the committee. Individuals organizing the fundraising identify guests that might include those who cannot be physically present for reasons of distance but send their financial contributions through money transfer systems. Under this system, qualification for assistance consists of the costs of hospitalization, payment of school fees, and paying fines, among other needs.

The system uses elaborate communication systems when organizing fundraising efforts. Neighbors and friends are informed about the upcoming contributions and pledges to be taken from them. Using modern communication technologies such as mobile phones, WhatsApp groups, e-mails, and Facebook, individuals—including those in the diaspora—can now be reached easily. Tapping into wide-ranging economic opportunities can raise vast sums of money. We use a case example of one neighborhood group in Sololo-Ramata as an illustration.

A Case Study of Neighborhood Network Performance

We used 11-year archival records of a neighborhood in the peri-urban area of Sololo-Ramata town, where a general practice is to organize public fundraising activities once four or more times a month. Using records of the number of participants and amounts of cash raised through public fundraising were used to assess the performance of the neighborhood social capital network (Fig. 8.1A and B). Fluctuations in sums of money raised usually indicate the economy's general performance, especially during drought years. Over 11 years, the average number of participants was 329, with the total cash raised estimated at 600,000 USD (1 USD = 100 KSh). The number of participants and the funds raised between 2002 and 2012 varied over time (Fig. 8.1B).

Two central problems have arisen in the neighborhood social capital network system. Firstly, the number of needy people—that is, in the total pool for the neighborhood (when translated to the broader Obbu region)—is much larger than what clans would have had to deal with in the past. This requires more frequent fundraising activities. Secondly, demands for addressing a multiplicity of societal needs create economic fatigue among participants. This is indicated by fluctuations in the number of participants, suggesting that collecting cash from the same neighborhood without a break might be causing a lethargic attitude among the

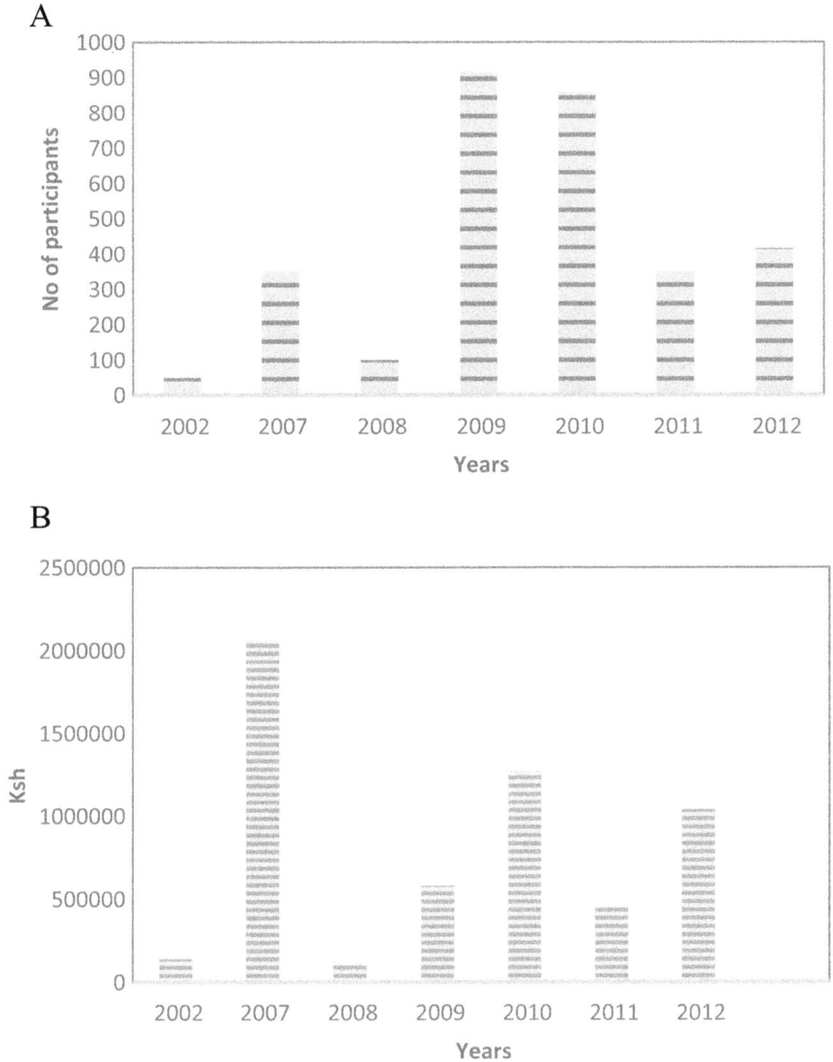

Fig. 8.1 Number of participants (A) and total cash (Ksh. 100 = 1 USD in 2008) raised (B) in fundraising initiatives by a neighborhood in the peri-urban area of Sololo-Ramata town between 2002 and 2012

population. The drought years 2002, 2008, and 2011 experienced fewer participants and lower accumulated funds. It would be interesting to consider how social capital networks provided the building blocks for transformation and resilience in pastoralism.

Building Resilience in Social Capital Networks

In the current study's context, the observed trend is like historical times (see Chap. 3). After experiencing pastoral collapse, the society reorganized its use of social capital networks. The first step was to build cooperative clan alliances, and the second was to build neighborhood social capital networks to consolidate the resources of all the clans that, in the past, had operated independently. Whichever way we examine the performance of social capital networks, their purpose is to organize and support the economic recovery of their members.[42] Like in the clan social capital networks, neighborhood actions are activated particularly after a crisis. The aim is for the society to use financial assets (as opposed to livestock assets) to support individuals in the neighborhood who are in dire need, regardless of their clan affinities. Formalizing neighborhood social capital networks allows members to draw on resources not directly available to the clan networks. Neighborhood social capital networks have had positive effects through their capacity to reorganize society. They could provide the mechanism in the long term for restoring the resilience of societal services and functions.

In this case study, the transformed social capital network does not draw on trans-frontier relations but has developed new vertical and horizontal structural ties[43] within and beyond neighborhoods. Its ability to utilize modern communication systems has developed capacities unavailable to clan social capital networks. The obligations for supporting the networks have changed from those of cooperative clans to the willingness of individuals to participate and contribute according to their capacities and wishes. In that way, the new neighborhood social capital networks are self-reinforcing and self-motivating, without requiring force (if necessary), as in clan social capital networks. Additionally, in the case of emergencies, the

[42] Aldrich, Daniel P. "Fixing recovery: Social capital in post-crisis resilience." *Journal of Homeland Security, Forthcoming* (2010).

[43] Bottrell 2009. Bottrell, Dorothy. 'Understanding "marginal" perspectives: Towards a social theory of resilience.' *Qualitative Social Work* 8.3 (2009): 321–339.

new social capital networks are more responsive and can be promptly organized to assist members in need. Yet, like the clan social capital network system, they function when there is trust and a feeling of obligation to support others.[44]

Accordingly, building community resilience is a collective action in which individuals provide financial help and make access to resources that may be widely available within and outside neighborhoods—such as involving the diaspora to support family needs and emergency programs associated with disasters. This means that strengthening neighborhood networks depends on building and maintaining social infrastructure that supports members during emergencies and, by so doing, attempting to restore resilience in the socio-economic system.[45] One may sense social stability when dealing with emergencies by assessing the bonding and bridging relationships within and across neighborhoods (i.e., the horizontal and vertical integration).

Neighborhood social capital networks' composition, needs, and demands are dynamic compared to clan social capital networks. This means that in these networks, the strengths lie in building bridges among localized groups to draw on external and internally located resources. While the actual impacts of the reciprocities are local, outwardly, the catchments from which resources are drawn reach far and wide. The neighborhood social capital network works on trust and obligations in which individual members are contributors and beneficiaries. A weakness is that frequent requests for fundraising contributions tend to cause fatigue among the public, resulting in lower levels of donations. Unlike the clan and individual social capital networks, the composition and needs of neighborhood networks might shift from time to time. The implication is that in neighborhood networks, the strengths lie in building bridges and relationships among localized groups to draw on internal and external resources. This is possible because the connections are highly embedded within the civil society and development programs, thus allowing neighborhood groups to enjoy resources outside the remit of traditional social capital networks.

Consequently, despite the neighborhood networks being inward-looking, through individual members, they functionally facilitate the

[44] Dynes, R. l. R, 'Community social capital as the primary basis for resilience,' (2005).

[45] Aldrich, Daniel P., and Michelle A. Meyer, 'Social capital and community resilience.' *American Behavioral Scientist* 59.2 (2015): 254–269.

building of networks within and across groups.[46] This is mainly because the relationships among the neighborhood social capital networks are highly embedded within civil society.[47] Also, development programs now allow neighborhood groups to enjoy resources outside the traditional functions of clan social capital networks. A more resourceful neighborhood social capital network would be more prepared to tackle emergencies and restore public confidence after a crisis than less imaginative neighborhoods. This being so, we are yet to test whether neighborhood social capital networks can deal with widespread social and economic collapses. Accordingly, we might learn lessons from how horizontal structures of neighborhood social capital networks operated by women function during regular times and in crises (Chap. 9).

[46] Pferfferbaum, B., Van Horn, R. L. and Pferfferbaum, R. L, 'A conceptual framework to enhance community resilience using social capital.'

[47] Dynes, Russell R., *Community social capital as the primary basis for resilience* (2005).

CHAPTER 9

Resilient Neighborhood Household Food Security: Women's Social Capital Networks, 1987–1996

This chapter focuses on neighborhood household food security operated by women in rural and peri-urban/urban environments.[1] This aspect of the moral economy needs to be visible to formal systems of food redistribution.[2] Our interest is in an informal local institution called *maaro*—meaning 'giving in turn'—among women in the Borana society. This sharing through social capital networks is not defined strictly in terms of financial relations but refers to sharing minute quantities of foods to build social solidarity.[3] While the amounts transferred are often minute, women's collective networks provide an effective adaptive strategy for maintaining the resilience of neighborhood household food security networks,[4] with the following characteristics: First, the networks can link the production, consumption, and marketing of food items to be distributed among

[1] Mamade, G. F., Womber, B. and Atlew, H., 'The role indigenous social institutions on pastoral and agro-pastoral households' access to and control over livelihood assets: Evidence from Somali region, Ethiopia,' *International Journal of Sciences: Basic and Applied Research*, 30 (5) (2016): 215–225.

[2] Scott, James C. *The moral economy of the peasant.* Yale University Press (Yale, 1977).

[3] Beaman, Anne W. 'Women's participation in pastoral economy: Income maximization among the Rendille.' *Nomadic peoples* (1983): 20–25.

[4] Amare A. and Simane, B., 'Does adaptation to climate change and variability provide household food security? Evidence from Muger sub-basin of the upper Blue-Nile, Ethiopia,' Ecological Processes (2018): 7: 13, https://doi.org/10.1186/S13717-018-0124x

G. Oba, *Pastoralist Resilience to Environmental Collapse in East Africa since 1500*, https://doi.org/10.1007/978-3-031-48291-5_9

households. Second, women use the income from the sales of dairy and grain products to purchase other food items, influencing food distribution through neighborhood household food security networks. Third, during droughts, when dairy and grains are scarce for marketing, the market prices of scarce foods interfere with food distribution.[5] Fourth, when available, integrating agropastoral produce with markets positively influences household food security by enabling credit facilities at food shops and kiosks.[6] The fifth is that quantities and monetized values of food shared by households differed according to wealth ranks.

The chapter comprises five sections. The first section presents a framework for gendered neighborhood household food security networks. The second characterizes women's social capital networks. The third describes the integration of rural and peri-urban social capital networks. The fourth evaluates quantities and monetized values of shared food, and the fifth discusses rebuilding the resilience of neighborhood household food security networks.

Gendered Neighborhood Household Food Security Networks

A framework for gendered social capital networks incorporates perspectives on assessing the effectiveness of social safety nets and risk sharing.[7] Structurally, gendered social capital networks involve cooperation by members of women's networks.[8] Food gifts are asked for when a household is in dire stress or may be unsolicited based on a history of food

[5] Maxwell, S. and Smith, M., *Household food security: A conceptual Review* (Part 1), 1992.

[6] Esenu, B., Ossiya, S., Oluka, J., Aliu, D., Owesigire, B., Ebiyau, G. and Olokojo, B., 'Positioning agro-pastoral women in livestock production: The link between socio-economic factors and improved household food security,' TROPICULTURA, 10th Anniversary of VSF—Belgium 2005.

[7] Endris, G. S., Kibwika, P., Obaa, B. B., and Hassan, J. Y., 'Does the implementation of social safety net intervention affect indigenous social capital systems for coping with livelihood shocks? Ethnographic evidence of agro-pastoral communities in Eastern Ethiopia,' in Efobi, U. T. and Asongu, S. (eds.) *Financing sustainable development in Africa*, 2018, https://doi.org/10.1007/978-3-3-319-78843-2_11

[8] Story, W. T., Tura, H., Rubin, J., Engidawork, B., Ahmed, A., Jundi, F., Iddosa, T. and Abrha, T. H., 'Social capital and disaster preparedness in Oromia, Ethiopia: An evaluation of the "women empowerment" approach,' *Social Science & Medicine* 257 (2020): 111907.

sharing. The giver expects the receiver to reciprocate in the future.[9] The frequency of giving food and other materials depends on the economic assets and capacities of the participating households.[10] The network mobilizes internal and external resources to be shared through household transfers to maintain the resilience of food security.

Ansah and co-workers[11] identify three attributes that promote the resilience of food security systems. First is absorptive capacity—that is, availability of resources, affordability, and consumptive power that reduces food insecurity risks. Second, households can make informed choices through appropriate adaptive actions to combat food insecurity. Third, through transformed capacity, families may adapt to changes in food supplies through necessary changes in livelihoods. Once these actions have been achieved, neighborhood household food security networks will maintain resilience.[12]

In resilient food security networks, innovative methods for acquiring new food items provide greater 'absorptive capacity' to individual households in buffering food insecurity. The implication is that, with increased absorptive capacity, families would be more able to sustain food security than before.[13] This might include building credit with food shops to expand facilities for obtaining extra resources that would not interfere with existing individual obligations, the social ethos of food sharing, and breaking food taboos. Individually, adaptive capacities might involve mechanisms for acquiring new food types that radically transform traditional food systems. The ability to do so varies from one household to another. Consequently, existing social capital networks might influence individual adaptive strategies—suggesting that wealthier and poorer

[9] Dahl, Gudrun. 'Women in pastoral production: some theoretical notes on roles and resources,' *Ethnos* 52.1–2 (1987): 246–279.

[10] Rakodi, C., 'A livelihood approach-conceptual issues and definitions,' in Rakodi, C. and Lloyd-Jone, T. (eds.) *Urban livelihoods: A people-centered approach to reducing poverty*, Earthscan Publ. Ltd., London, 2002.

[11] Ansah, G. K., Gardebroek, C. and Ihle, R., 'Resilience and household food security: A review of concepts, methodological approaches and empirical evidence,' *Food Security* 11 (2019): 187–1203.

[12] Ado, A. M., Savadogo, P., Abdoul-Azize, H. T., 'Livelihood strategies and household resilience to food insecurity: Insight from a farming community in Aguie district of Niger,' *Agriculture and Human Values* 36 (2019): 747–761.

[13] Ibid.

households might disclose varied capacities for food sharing.[14] Conversely, transformation capacities increase the uptake of new techniques, such as combining informal and formal networks and linkages to markets for food supplies and credit-client establishment. Neighborhood and distant networks are linked up through bonding and bridging relationships.[15]

Bonding and Bridging Relationships

Woolcock[16] describes different kinds of connections within horizontal and vertical networks. Horizontal networks are the connections between proximate members, while vertical networks are connections between distant or external members for sharing resources over time.[17] In neighborhood networks, sharing is through bonding (i.e., horizontal networks), while remote network members (i.e., vertical networks) are connected through bridging relationships. Bridging often involves vertical networks, including formal institutions that provide resources women use to strengthen their informal networks. Bridging relationships increases absorptive capacities among distinct communities, such as rural and peri-urban communities,[18] thus connecting households across geographical spaces that cannot be accessed through social bonding. Through a bridging relation between individuals and groups of other economic status—such as shopkeepers and clients—debt-clientship is created with individual women.[19] We will use this background to discuss the characteristic features of neighborhood food security networks among the Obbu Borana households.

[14] Galvin, K. A., Thorton, P. K., Boone, R. B. and Sunderland, J., 'Climate variability and impacts on east African livestock herders: The Maasai of Ngorongoro conservation area, Tanzania,' *African Journal of Range and Forage Science* 21 (3) (2004): 183–189.

[15] Endris, G. S., Kibwika, P., Obaa, B. B., and Hassan, J. Y., 'Does the implementation of social safety net intervention affect indigenous social capital systems.'

[16] Woolcock, M., 'The place of social capital in understanding social and economic outcomes,' *Canadian Journal of Policy Research* 2 (1) (2001): 11–17.

[17] Dapilah, F., Nielsen, J. Ø. and Friis, C., 'The role of social networks in building adaptive capacity and resilience to climate change: A case study from northern Ghana,' *Climate and Development* 12 (1) 2020): 42–56.

[18] Ibid. Blackburn, Morgan, Alejandro Guzman, Jeff Lieberman, and Anne Sprinkel. "Assessing the role of social capital in agro-pastoral resilience in the Sahel: A systems Perspective." *Coordination with Mercy Corps and the Elliott School of International Affairs, George Washington University* (2014).

[19] Endris, G. S., Kibwika, P., Obaa, B. B., and Hassan, J. Y., 'Does the implementation of social safety net intervention affect indigenous social capital systems.'

Characteristic Features of Neighborhood Households' Food Security Networks

From discussions with women randomly sampled across Obbu, in 1995, 97.7 percent share *maaro* relationships with neighborhood and distant households, while 2.2 percent do not. The networks allow members to share resources daily and occasionally through bonding and bridging connections. Among those who share *maaro* networks, only 3.3 percent have single partners, while 96.7 percent involve multiple partners. In multiple partner relations, women's social capital networks create webs of social and economic ties across the community.[20] We use narratives of women to illustrate these relationships in this section.

Women's Narratives

Society has mechanisms for creating new relationships, as friends and relatives are often widely spread. Lokho Waario had one such experience: 'We are new in this village. We moved from our former settlement and settled here for security reasons. I have…begun sharing a few things with my immediate neighbor.' The quantities shared might be minute and for emergencies.

Lokho Roba stated, 'Sometime in the morning, when a herder is leaving with the cattle and tea leaves are not in the house, I visit my neighbor. I cannot visit the town for that small quantity that early. My neighbor would give me 2 to 3 spoonsful, or a palmful of tea leaves, enough for one preparation.' Another example of sharing is collecting maize from a neighbor's grain store if a household has exhausted its own, as described by Dakki Happicha: 'If it is the maize we harvest from our farm, we may share it with neighbors and relatives, who failed to harvest any or used up their own.'

The Obbu women consider such relations to vary, being either strong, weak, or occasional. Sukule Boru explains: 'For those I share strong *maaro* relations with when I visit, I would be obliged to bring along items such

[20] Berzborn, S. and Schnegg, M. 'Vulnerability, social networks, and resilience: Rural livelihoods in the Richtersveld, South Africa, Research Gate,' https://www.researchgate.net/publication/309616695

as sugar, tea leaves, and firewood.' Halima Roba observed, 'You help your next neighbor with what you must to establish relations of giving and receiving. Today you give, and tomorrow you receive...'

Poor families are not known to have wide-ranging networks, so they might be hesitant to visit their neighbors for food sharing. Such families suffer risks of severe food shortages. Therefore, it becomes the neighbors' responsibility to send them food. In such arrangements, the household woman usually requests their husband to help the neighbor. This was the case of the family of Diraamu Gaayya—a widow and poor. She narrated how she received help from a neighbor: 'We did not receive any harvest of maize this year [1995]. The maize I am cooking was given to us by our neighbor. I never begged. They sent me the grain because they knew we have no food.' We inquired how women with disabilities might be involved in social networks. Qallu Qax'u is blind, but this did not end her participation in *maaro*. She stated, 'I am a blind woman, and I give what I have to my neighbors because my survival depends on them.' Women depend on relief organizations for food, which they redistribute through social networks. For example, some households we interviewed claimed they shared social relationships with the Catholic Mission at Sololo–Ramata. We will examine the cases of five families.

Individual Social Networks Shared with a Relief Organization

Individual women—the majority of whom were converts and regular attendees at church services—used the Mission as part of their social capital networks. Partnerships between individual households and the Catholic Mission were developed through regular contact. The Mission personnel (usually the nuns) got to know the needy families. Rather than providing free goods and services, mutual agreements were made that required disadvantaged households to supply the mission with the labor needed or other materials, such as firewood. The homes, in return, received food and the opportunity to sell the products of women members, such as chicken and eggs (Table 9.1). At the Sololo Mission Hospital, individual households were organized to supply eggs and chicken, which became a regular source of cash to purchase other foodstuffs. The activity with the highest level of participation was drinking water from the borehole of the Mission Hospital during the dry season, firewood supplied to the mission in

Table 9.1 Social networks established with the Catholic Mission (N = 115)

Actions	*% Household participation*
Water supply	24.4
Casual jobs	8.8
Cash gifts	1.7
Medical assistance	3.4
Children's education	13.9
Supplying firewood	13.0
Supplying chicken and eggs	1.7
Employment	10.5
Non-participants	8.7

exchange for food, and famine relief became part of shared networks with the Mission (Table 9.1).

Local households had demonstrated a solid adaptive capacity to transform their traditional food sources. Another innovative food source was obtained by participating in the ownership and management of poultry, which, although common among sedentary households, was never part of traditional pastoralists' livelihoods. The women stated that acceptance of poultry by Borana households is a new adaptive strategy. The first reason was that chicken was introduced into homes by school-going children. The second reason was that women find it easier to manage poultry under the settled lifestyle. The third reason was that chicken is a means of earning quick cash income. The money is then used to buy school children food, books, pens, and uniforms. The fourth reason was that poultry was among the means individual women used to build food-sharing networks with the Catholic Mission and the Mission Hospital. We will present the five case families.

Case Families

The household of Kula Ali is a typical one that participates in food-sharing networks among immediate neighbors and the Catholic Mission. Kula is a widow with a family of nine children. Two of her sons have completed secondary-level education, and one daughter is employed at the Post Office. Three of her sons and four of her daughters are married and running independent homes. Her daughters and daughters-in-law form her immediate social networks. She also shares food with eight other neighbors. Her working daughter remits money at the end of every month, with which Kula Ali pays her debts and purchases food—to be shared with friends. Before the drought of 1991/92, Kula Ali had robust social capital networks. After the drought, most relations, especially with distant partners, were rendered inactive. This is partly because her resources had diminished. She sold eggs to the nuns at the Mission Hospital and received regular supplies of famine relief from the Catholic Mission, which sustained her food security during times of scarcity.

The second family is Sukule Boru, a widow with three children. She had dynamic social capital networks in the neighborhood and distant places. Her son was working at Kenya Utali College. He sent regular remittances for her to buy food and pay debts at food shops. She managed livestock and conducted farming activities. Among her social capital partners are her stepmother, grandmother, and non-relative. Their networks involve the distribution of sugar, tea leaves, firewood, water, etc. She also shares a social network with distant households. She presents them with tea leaves, milk, and water when she visits. In return, they provide gifts when they see her. Among her distant social capital networks, members are those residing across the frontier in Ethiopia, with whom she often shares relief food.

The third household is that of the Hakule Waqo, representing an enterprising but poor household. Her exceptional resourcefulness was her ability to build a vast social network, including non-traditional associates. She considered the Catholic Mission nuns—part of her social network, with whom she shared resources, but under different rules and regulations. Among her nine social network partners, about 33 percent comprised relatives, and the rest were non-relatives. She considered some of them reliable partners and occasionally shared food with others. Her relationship with the Catholic Mission was strengthened by her membership in the church and the friendship relationship she established with the nuns. During the 1991/92 drought, the Mission supplied her with domestic water. She sold eggs and periodically chicken to the nuns. At other times, she was given a contract to supply firewood to the Mission Hospital and, in return, received food, which she shared through her networks. Her other sources of income were from the sale of poultry and eggs.

The fourth household is that of Tarri Deereva, which represents the poorest of the poor. She had eight chickens and 11 goats. She shared social networks with two other homes, both of whom were poor. She relied on the Catholic Mission for food security; even so, her family frequently went hungry. She had a different experience with the missionaries, claiming she had fallen out with them since they provided insufficient food for her family. Therefore, she and her family were in a perpetual food insecurity situation.

The fifth household is Diiramu Jaldeesa, who was relatively wealthy before the drought of 1991/92 and enjoyed sustainable food security. The family had a large flock of chickens but consumed most of them during the drought. Diiramu shared social networks within the neighborhood and with distant associates. Her relationships with the mission were occasional when she needed food. A far greater adaptive strategy by the Obbu Borana women was the integration of the rural and peri-urban social capital networks.

Integration of Rural and Peri-urban Social Capital Networks

In rural (pastoral) and peri-urban (agropastoral) contexts, the relationships women build are dynamic and multi-layered due to their varied services.[21] Thus, rural and peri-urban households are independent entities linked through market forces. Socio-economic relations might be influenced by migrations and the exchange of food materials as part of women's social networks across the rural and peri-urban continuum.[22] Visits to rural and peri-urban/urban areas provide the resources households operating in different geographical spaces need.[23] The relationships are characterized by the movement of people and goods to mitigate risks associated with resource shortages.[24] Rural households have strong family networks, while

[21] Meikle, S., 'The urban context and poor people,' in Rakodi, C. and Lloyd-Jone, T. (eds.) *Urban livelihoods: A people-centred approach to reducing poverty*, Earthscan Publ. Ltd., London, 2002.

[22] Laquinta, D. L. and Drescher, A. W., 'Defining peri-urban: Rural-urban linkages and institutional connections,' ResearchGate Nebraska Wesleyan University, http://www.researchgate.net/publication/287613842

[23] Owuor, S. O., *Urban households ruralizing their livelihoods: The changing nature of urban-rural linkages in an East African town*, Paper presented at the African Studies Centre Seminar series, Leiden, Thursday 16 December 2004.

[24] Jamshed, A., Birkmann, J., Feldmeyer, D. and Rana, I. A., 'A conceptual framework to understand the dynamics of rural-urban linkages for rural vulnerability,' *Sustainability* 12 (2020): 2894, https://doi.org/10.33010/su/2072894.

peri-urban residents have strong neighborhood networks.[25] In a traditional rural neighborhood, individuals share spatial settings. This may be contrasted with sedentary peri-urban settlements where families live on physically isolated plots. Whereas in rural communities, building friendships starts inside the huts, the peri-urban areas are exclusive spaces where the extent of going in and out is limited. Thus, the spatial neighborhood influences the relationships that might be established, affecting resource flows.

Women in peri-urban areas built more secure neighborhood social capital networks than those from rural households with fewer partners in town (Table 9.2). Accordingly, in highly transformed socio-economic systems,

Table 9.2 Frequencies (%) of participation in maaro networks among rural and peri-urban households

Criteria	*Rural networks*	*Peri-urban networks*
Location of network		
Neighborhood	63.1	90.6
Same settlement	94.7	100
Neighboring settlement	5.3	31.2
Town	5.3	18.7
Shared items		
Food resources	84.2	90.6
Economic support	26.3	12.5
Social support	73.7	93.7
Personal services	31.6	78.1
Frequency of participation		
Every day	–	62.5
Occasional	42.1	43.7
Frequent	52.6	9.4
Season of most activity		
Wet season only	–	–
Dry season	36.8	21.9
Drought year	36.8	34.4
All the time	52.6	56.2
Problems in acquiring food		
Lack of money	100	90.6
High food prices	63.1	83.4
Problems of availability	94.7	96.9

[25] Wells, M., 'Resilience in older adults living in rural, suburban, and urban areas,' *Journal of rural Nursing and Health Care* 10 (2) (2010): 45–54.

such as peri-urban/urban areas, the materials exchanged and the network participants might differ from those in rural environments. In particular, the advantage for peri-urban households is that they have access to a broader range of sources into which they may tap to access food and other services, for example, the socio-economic relations women share with shops and other food retail facilities. Accordingly, peri-urban households are more actively involved in the daily and occasional sharing of foods and services. At the same time, in the rural women's networks, participation is occasional or frequent (not every day) (Table 9.2). This might suggest that peri-urban women rely more on their networks to access food than rural women. We were also interested in understanding the seasons of the year when most resource sharing occurs and found that it takes place all the time (i.e., wet and dry). In rural and peri-urban women's social networks, drought years are periods of lower levels of food exchanges—indicative of general scarcities.

Among the households interviewed, problems in accessing food were common, arising from the need for more money, high food prices, and food availability. These challenges have been overcome by combining informal and formal social networks for resource sharing.

Informal and Formal Women's Social Networks

Participation in collective women's groups (i.e., formal networks) is a recent development among the Obbu households. Using public fundraising opportunities, women of the same settlement join, registering as groups that are legal entities. New socio-economic options are available to improve the welfare and empowerment of women through collective action.[26] These formal social networks often comprise members from informal networks. Further, the same household might participate in more than one group to diversify their sources of income and livelihoods.[27]

The formal women's groups are led by women and are aimed chiefly at accessing funding from government departments, politicians, and

[26] Kibet, S., Nyangito, M., MacOpiyo, L. and Kefach, D., 'Tracing innovation pathways in the management of natural and social capital on Laikipia Maasai Group ranches, Kenya,' *Pastoralism: Research, Policy and Practice* 6 (2016): 16, https://doi.org/10.1186/s13570-016-006-32.

[27] Dapilah, F., Nielsen, J. Ø. and Friis, C., 'The role of social networks in building adaptive capacity and resilience to climate change: A case study from northern Ghana,' *Climate and Development* 12 (1) (2020): 42–56.

development agents to improve entitlements for their members.[28] Thus, while informal social networks rely on internally produced resources, formal women's groups depend on externally produced resources.[29] Traditional women's organizations promote income activities so that women may use the funds to provide micro-credit for their members.[30] In turn, the members use the income to strengthen their informal networks.[31]

We found that more women in rural areas participate in the formal system than in the peri-urban areas. This is partly due to the greater attention they received from development agencies. Using their numbers to their advantage, women's groups become active agents of development changes.[32] The groups often mobilize their efforts in helping each other during social engagements, such as weddings, burials, and other social occasions. Women's empowerment, through the organization of groups, has therefore created higher levels of social cohesion (Table 9.3). The

Table 9.3 Comparing participation (%) in formal women's networks in rural and peri-urban communities in the Obbu region

Survey questions	*Rural networks*	*Peri-urban networks*
Participation in formal networks (%)		
Percentage of households	84	58
Development needs	50.0	66.7
Social engagement	32.1	90.0
Economic support	28.1	53.3

[28] Ngoitiko, M., *The pastoral women's council empowerment for Tanzania's Maasai: The role of local organisations in sustainable development*, id Gatekeeper 137e, August 2008.

[29] Khalif, Zeinabu Kabale, and Gufu Oba. "Continuity and Change in Social Networks among Borana Women in Northern Kenya." *Nomadic Peoples* 22.2 (2018): 249–281.

[30] Coppock, D. L., Desta, S., Tezera, S. and Gebru, G., 'Capacity building helps pastoral women transform impoverished communities in Ethiopia,' *Science* 334 (2011): 1394–1397.

[31] Njoki, K. R., Assa, O. C. and Mulama, D., 'Strategies used by women to adapt to the effects of climate change in Mwala Division, Machakos County, Kenya,' in Khalif, Z., Gachene, C. K.K., Gicheru, P., Mburu, D. M. and Gakahu, C. G., *Sustainable land management in dry lands of Kenya: Improving land productivity through participatory Research and Technology transfer*, UNDP, Nairobi Kenya, 2015.

[32] Story, W. T., Tura, H., Rubin, J., Engidawork, B., Ahmed, A., Jundi, F., Iddosa, T. and Abrha, T. H., 'Social capital and disaster preparedness in Oromia, Ethiopia: An evaluation of the "women empowerment" approach,' *social Science & Medicine* 257 (2020): 111907.

growth of formal women's groups is a response to increased socio-economic needs.[33] In the next section, we consider the role played by marketing dairy products and grain sales in social capital networks.

Links Between Market Supplies of Agropastoral Products

Dairy producers adopt two marketing strategies: (a) sell milk on the open market and (b) engage customers billed at the end of each month.[34] Among the members of women's networks, those in the rural site participated more in milk sales than the peri-urban households. This is partly explained by more significant livestock populations in the rural than in the peri-urban area. Homes with insufficient milk stock sell their produce on the open market. Individuals with more milking cows tend to maintain prices at fixed rates by billing their regular customers. The cost of milk during the wet season tends to be lower due to a glut in the market, compared to the dry season when milk prices increase due to scarcity (Table 9.4).

The rural households supply milk mostly from cattle, while the peri-urban households, primarily consumers, rely on both cattle and camel milk. Camel dairy products have become an important trade commodity in peri-urban areas, following growing demand for camel's milk. In northern Kenya, more significant volumes of camel milk are sent to the market to supply peri-urban residents.[35] The rural households that produce surplus milk sell it to the peri-urban and urban residents, serving as clients who are billed at the end of each month. Milk bills are an essential income source for rural and peri-urban households. In 1987, the peri-urban households earned KSh 900 per month, while the rural households earned

[33] Khalif, Z. K. and Oba, G., 'Continuity and change in social networks among Borana women in northern Kenya.'

[34] Oba, G., 'Changing property rights among settling pastoralists: an adaptive strategy to declining pastoral resources.' pp. 38–44. in: P.T.W. Baxter & R. Hogg (eds.). *Property, poverty and people: changing rights in property and problems of pastoral development*. Manchester, UK Department of Social anthropology and /international Development Centre, University of Manchester 1990.

[35] Anderson, D. M., Elliott, H., Kochore, H. H. and Lochery, E., 'Camel herders, middle women, and urban milk bars: The commodification of camel milk in Kenya,' *Journal of Eastern African Studies* 6 (3) (2012): 383–404.

Table 9.4 Milk marketing by rural and peri-urban households

Survey questions	*Rural networks*	*Peri-urban networks*
Participants in milk market (%)	26.3	15.6
Prices (KSh) in milk/liter		
Wet season	8.08± 2.38	8.67 ± 2.71
Dry season	12.61± 2.40	12.92 ± 2.48
Source of the milk (%)		
Cattle	68.4	43.7
Camels	–	6.2
Both	–	12.5
Number of people practicing milk bills (%)	15.8	9.4
Average earnings from milk bills in a month (shillings)	450 ± 50.1	900
Average number of months for milk bills	8.67 ± 8.08	8
Ways milk bills help households		
Sustain livelihood	–	3.3
Serve as collateral for credits	5.3	6.2
Purchase foods and other goods	10.5	9.4

half (Table 9.4). The income is a collateral source for accessing credit at shops and food kiosks and purchasing non-pastoral foodstuffs.

Among 164 households surveyed in 1987, 79 percent maintained milk clients before the drought of 1991/92; after that, only 25 percent of the households engaged in milk sales maintained the transactions. After the drought, 75 percent of the households reported non-participation in the milk trade. The changes resulted from high cattle losses and the slow growth of milk herds (see Chap. 7).

The number of meals households serve daily, and the amount of grain consumed are food security indicators. About 13 percent of the households reported having three meals daily, including breakfast. The changes reflect how the sedentary lifestyle has transformed society's consumptive food patterns to become comparable to urban residents. The amount of grain intake had also increased per household, with the total volume of grain consumed dependent on family size. On average, households consumed about 2 kg of grain per day; those that consumed 1 kg per meal (comprising 19.5% of the total households) were probably poor or had fewer members. The annual grain requirement of an average household was 1440 kg. The increased dependence on grain markets means that the

Obbu households are vulnerable to supply levels and price fluctuations.[36] According to the informants, lack of cash, high food prices, and food scarcity are risk factors that affect food sharing among neighborhood households. Rural and peri-urban households enjoyed credit facilities at the shops and food kiosks.[37]

Credit Facilities Enjoyed By Individual Households

The peri-urban and urban milieux offer additional resources to women. Women in peri-urban establish relationships with retail shops where cash transactions are used to access credit facilities.[38] Considering that urban livelihoods are highly commercialized, households are obliged to seek additional sources of income to acquire necessary resources. The capacity of peri-urban households to meet their social and economic goals demands that they have access to cash from such sources as milk bills and grain sales to cushion their consumptive needs.[39]

The shopkeepers allow clients to take sugar, tea leaves, cooking fat, and other consumable and non-consumable goods on credit. A general agreement is established with a retail business on when and how to pay any debts owed. For the business, the clients provide a regular pool of customers, some of whom also use the business to deposit extra cash that can be used later to cover purchases, with any outstanding balance to be paid at an agreed time. Such regular customers, most of whom also have established credit facilities, provide security in a non-competitive business. For the customers, the credit facilities serve as security against food shortages when their sources diminish.

Our survey data confirms that many women in rural and peri-urban households can access credit facilities at local shops (Table 9.5). The

[36] Helland, J. 'The role of sharing in Borana pastoralism'. Paper presented at a workshop on 'Perspectives in survival strategies in African arid lands,' Organized by the Scandinavian Institute of African Studies, Aarhus, Denamark, 3–6 October 1991.

[37] Badzaban, F., Rezaei-Moghaddanam, K. and Fatemi, M., 'Analyzing the individual entrepreneurial resilience of rural women,' *Journal of Global Entrepreneurship Research*, https://doi.org/10.1007/s40497-021-00290-1

[38] Fratkin, E. and Smith, K., 'Women's changing economic roles with pastoral sedentarization: Varying strategies in alternate Rendille communities,' in *As pastoralists settler*, Springer, Boston, M. A., 2005.

[39] Rakodi, C., 'A livelihood approach-conceptual issues and definitions,' in Rakodi, C. and Lloyd-Jone, T. (eds.) *Urban livelihoods: A people-centred approach to reducing poverty*, Earthscan Publ. Ltd., London, 2002.

Table 9.5 Household accessibility to credit (%) at shops, in support of social networks

Survey questions	*Rural networks*	*Peri-urban networks*
% people who have credit facilities at local shops	89.5	90.6
Providers of credit facilities		
Shops	84.2	96.9
Relatives	5.3	–
Neighbors	–	6.2
Others	–	–
Types of credit facilities		
Grains	68.4	93.7
Commercial food	78.9	100
Money	–	28.2
School fees	–	21.9
Clothes	5.3	28.2
Conditions for granting credit		
Agreement with clients	42.1	46.9
Guarantor	15.8	25.0
Property as collateral	31.6	28.2
Past creditworthiness	5.3	9.4
Seasons when credits are needed		
Wet season	5.3	43.7
Dry season	15.8	3.1
Both	–	37.2
Drought year	21.0	37.4
All the time	52.6	50.0

credits are mainly grain, commercial foods, and non-food items. Individual women negotiate credit facilities to be paid back at an agreed time. The criteria for establishing credit facilities are agreements with shopkeepers, on the recommendation of a guarantor, using the property as collateral, or past creditworthiness at the same business. Only peri-urban households rely on money deposits in cash and then take goods, when necessary, against that credit balance at shops. Whereas credits are required throughout the year, more credit facilities are needed during dry and drought seasons. Due to their proximity to markets, peri-urban households tend to be involved in petty trade and have the prospect of borrowing cash from friends and retail businesses. For example, creditworthiness can be established by those women who run little transactions and other small enterprises.[40] They, too, can be the source of credit to their customers. Another type of credit is a cash loan based on a prior arrangement between the lender and the borrower. The loan terms are verbally agreed upon, and the payment period is specified.[41] Such transactions are not limited to women's social networks but can also involve men, who guarantee how the loan will be paid. Nonetheless, droughts tend to disrupt debt payments. If the repayment of debts were to be delayed due to environmental shocks such as droughts, the households concerned would be put in perpetual liability. We quantified and monetarized domestically shared food items.

Quantities and Monetarized Values of Shared Food

The weekly expenditure and sharing by nine households—representing poor, medium-level, and rich wealth ranks based on their livestock holdings and amount of grain in their stores—in Golole were monitored from January to July 1987. Weekly household budgets were recorded based on actual purchases and monetary values of food exchanged through women's food-sharing networks. The survey was repeated thrice weekly, and the women requested to recall quantities of food received from friends and amounts purchased. The items transferred from one household to another, not consumed on the spot, were 'calibrated' and expressed in monetary values. The value of food shared compared with food purchased and what was produced measures an individual household's food security situation. It was assumed that the families were part of wider intra-communal food security networks.

The proportions (percentage) of total cash expenditure by each household rank on the five common foodstuffs and miscellaneous items (Fig. 9.1). The proportions spent on essential food items like oil, tea leaves, sugar, and milk are comparable for all wealth ranks. Poor households spend the most on purchasing clothes for their families (26.9% of their total spending). The amount of community development contributions charged to each household was commensurate with their economic

[40] Okello, G. A., and Ngala, O. M., 'Coping strategies among pastoralists women entrepreneurs against the odds: Experiences from Baringo Division, Samburu North, Kenya,' *International Journal of Business and Management* 12 (10) (2017): 166–176.

[41] Baird, Timothy D., and Clark L. Gray. "Livelihood diversification and shifting social networks of exchange: a social network transition?" *World development* 60 (2014): 14–30.

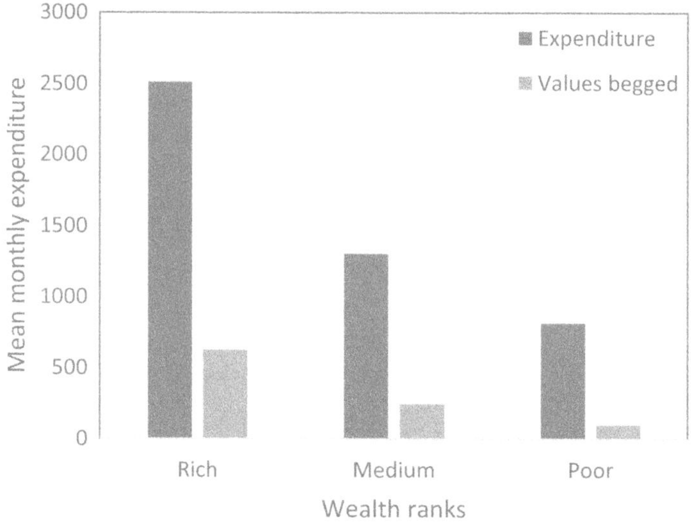

Fig. 9.1 Mean monthly expenditure and cash values of food items received through intra-communal sharing by rich, medium-level, and poor households in Golole. All values are in Kenyan shillings (1 US $=KSh 38 in 1987)

abilities. Thus, the wealthy and medium wealth households spent more on community development contributions than the poor households. Given that the money was often contributed to national projects from which the local communities derived no economic benefits, they considered this service as playing no role in food security.

The medium wealth households solicited a higher percentage of their expenditure on grain (11.9%) than the rich (5.5%) and the poor households (3.7%) combined (Fig. 9.2). Generally, amounts of grain solicited were insignificant compared to tea leaves and sugar. The frequency of requesting miscellaneous food items by the medium wealth and poor households is similar. The frequency of food solicited increases for all wealth ranks during ceremonial occasions when friends and neighbors provide additional milk and other food supplies. Intra-communal sharing also increases substantially during a child naming ceremony (*guubisa*), which is held in June when the maize has ripened, and milk supplies are usually plentiful. The traditional practice is for the households preparing

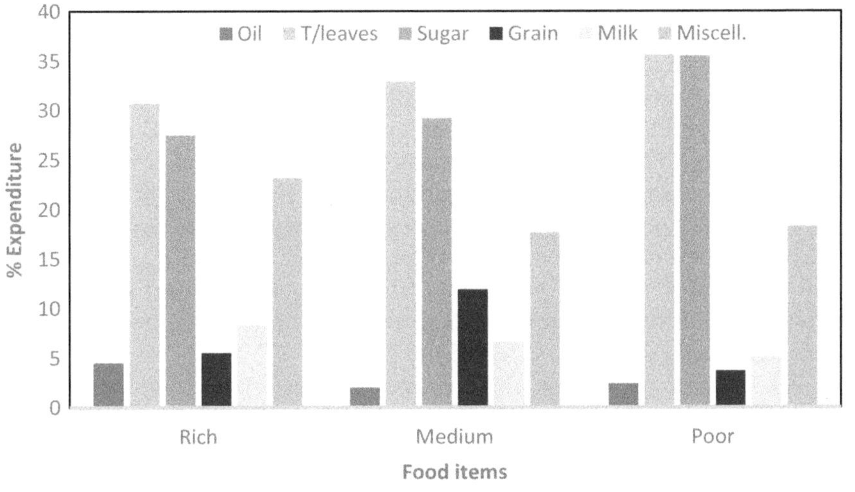

Fig. 9.2 Frequency of solicited common food items by wealth rank

for the celebration to request all friends, nearby and neighborhood members to prepare yogurt and donate other food items as part of their contribution (*gummata*). The consumption of illicit liquor increases during a child naming ceremony and is used as a substitute for milk, irrespective of the wealth rank of the households. As much as 25 liters of alcohol can be consumed at one celebration organized by a household of medium wealth. This habit is a break from tradition when ceremonies were considered cultural festivities.

The wealthy households had the highest mean monthly expenditure on common food items and the highest 'soliciting capacity' relative to the medium wealth and poor families. Among the rich homes, values of food solicited were equivalent to 24.9 percent of their total expenditure, close to the medium wealth and the poor households, where matters of food demanded were 18.9 percent and 12 percent of monthly spending, respectively. Across the wealth ranks, comparable percentage expenditures on tea leaves and sugar were recorded. We used the results to reflect on building resilience in neighborhood household food security networks.

Rebuilding Resilience of Neighborhood Household Food Security Networks

We examined the processes of rebuilding resilient food security in four contexts. The first context is the characteristic of social capital networks—the networks shared with members through bonding and bridging relationships. Women recognized the advantages of the proximate and distant networks. The neighborhood food security networks were personal, with women sharing minor food items when occasions demanded. Also significant is the capacity of the women to establish social networks shared with relief organizations. Using these external opportunities, women increased their abilities to secure food security.

The second context is the linkages between rural and peri-urban household social networks. The operation of social capital networks suggests that rural and peri-urban societies are embedded in webs of social relations of food sharing that are culturally determined.[42] The connectivity between rural and peri-urban households—through regular food and service exchanges—has enabled the system to remain resilient.[43] The sustained social ties highlight the diversification of livelihoods and spatial distribution of resources among members embedded in multiple networks in both environments.[44] Additionally, women took advantage of collective action by combining the resourcefulness of informal and formal systems.[45] The capacity of women to connect informal and formal networks transformed and expanded their capabilities for sharing resources

[42] Rockenbauch, T. and Sakdapolrak, P., 'Social networks and resilience of rural communities in the Global south: A critical Review and conceptual reflections,' *Ecology & Society* 22 (1): 10. https://doi.org/10.57511/Es-09009-220110

[43] Cassidy, L. and Barnes, G. D., 'Understanding household connectivity and resilience in marginal rural communities through social network analysis in the village of Habu, Botswana,' *Ecology & Society* 17 (4) (2021): 11, https://doi.org/10.5751/ES-04963-170411

[44] Bushyby, K. and Sites, E., *Resilience and risk in pastoralist areas: Recent trends in diversified and alternative livelihoods, Karamoja, Uganda* in USAID, 2016.

[45] Badejo, A. F., Majekodunmi, A. O., Kingsley, P., Smith, J. and Welburn, S. C., 'The impact of self-help groups in pastoral women's empowerment and agency: A study in Nigeria,' *Pastoralism, Research, Policy & Practice* 7 (2017): 28, https://doi.org/10.1186/s13570-017-0101-5.

during different seasons. The transformation does not negate the connectedness of the networks for coping with shocks associated with disasters. Through behavioral changes, it has been possible for pastoral women to transform social capital networks—a process that has maintained social resilience.[46]

The third context is the market-supply chain and using cash to buy non-agropastoral foods and links with the debt-client relationship at the food shops and kiosks. New adaptive strategies in women's social capital networks, such as dairy supplies and milk bills, and improved access to credit facilities for rural and peri-urban households—combined with individual entrepreneurship activities—increased the resilience of social capital networks.[47] The increased importance of collective action and access to credit facilities serve as new mitigating strategies for maintaining stability in social capital networks.[48]

The fourth context is the quantities of food items shared and the monetarization of food solicited. While social obligations for distributing food are fixed by social ethos and customs, food production and consumption might not relate directly to shared food. This is because the action of sharing food arises from personal desires unrelated to affordability. This would explain why food sharing reflects mutually reciprocal behavior, as giving has no strict rules other than personal decisions to help others.[49] The goal of long-term resilience in food security requires planning for the future besides maintaining immediate food status levels. The emphasis must be on building safety nets and adopting risk-reduction strategies. Chapter 10 summarizes key emerging issues on innovative adaptive systems from the case studies.

[46] Fratkin, E., 'Stability and resilience in East African pastoralism: The Rendille and the Ariaal of Northern Kenya' *Human Ecology* 14 (3) (1986): 264–286.

[47] Badzaban, F., Rezaei, Moghaddam, K., and Fatemi, M., 'Analysing the individual entrepreneurial resilience of rural women,' *Journal of Global Entrepreneurship Research*, 2021, https://doi.org/10.1007/s40497-021-00290-1

[48] Arjjumend, H., 'Review of pastoralists' resilience and adaptation to climate change: Can technology help pastoralists mitigate the risks?' *International Journal of Bio-resource and stress Management* 9 (1) (2018): 118–128.

[49] Béné, Christophe, Derek Headey, Lawrence Haddad, and Klaus von Grebmer's. Is resilience a useful concept in the context of food security and nutrition programmes? Some conceptual and practical considerations.' *Food Security* 8, no. 1 (2016): 123–138.

CHAPTER 10

Innovating Pastoral Resilience in the Future: A Synthesis

As the writing of this work was coming to completion in 2022, a regional drought broke out in southern Ethiopia and Kenya. The 2019–2022 droughts resulted in mass mortality of livestock, which for southern Ethiopia and northern Kenya is estimated at between 2.5 to 3.0 million heads.[1] The droughts disrupted the recovery of pastoralism following the famines of 2011 and 2015. From decadal droughts, we showed that up to 24 percent of the Obbu Borana households are likely to have dropped out of pastoralism (see Chap. 7). For the rest, the economy has been transformed through innovative adaptive strategies[2] through individual and collective actions in response to economic challenges provoked by the droughts. The reality is that all the innovation strategies described herein are by the society itself. The innovations have social and technical dimensions in terms of experimentation with new ideas to promote the resilience of the pastoral economy and in pursuit of new innovative adaptive

[1] This is reported in varieties of media outlets and local relief agencies.

[2] Scoones, I. and Adwera, A., *Pastoral innovation systems: Perspectives from Ethiopia and Kenya*, Future Agricultures, DFID, FAC Occasional Paper ST101, IDS, UK, 2009.

G. Oba, *Pastoralist Resilience to Environmental Collapse in East Africa since 1500*, https://doi.org/10.1007/978-3-031-48291-5_10

strategies.[3] Gaps remain in the narratives. In the future, pastoralists need education and training to improve their adaptive systems.[4]

The chapter has three aims. First, to synthesize the key findings from the case studies. Second, discuss progressive, innovative strategies that emerged from different case studies. Third, develop a rationale for field school and lifelong education training for the pastoralists on innovative strategies for mitigating future droughts.

A Synthesis: Connecting the Past to the Present

We will attempt to connect the past and present by summarizing the findings of the work organized around the three study thematic areas. These are described in the following sub-sections.

Regional Memories of Disasters

The work introduces characteristic features of socio-ecological systems and the resilience to a catastrophic collapse in pastoralism in the trans-frontier region of southern Ethiopia and northern Kenya by tracking pastoralists' disaster adaptations during three historical periods. The study defined terms such as adaptation, collapse, resilience, and transformation in pastoralists' disaster adaptive strategies. Patterns and cycles of triggers that drove disaster processes in the past enable us to recognize evolving adaptive systems, which help assess and evaluate resilience, collapse, reorganization, and transformation of pastoralism. By locating the study in historical and contemporary periods, the work outlines the processes that link the present to the past. Regional (historical) and local case studies are combined to map pastoralists' adaptive strategies trajectories over centuries (Chap. 1).

Analyses of historical research methods combined elements of social science, history, political ecology, and natural ecological systems (Chap. 2). Methods for measuring matrices of resilience, collapse, and transformations were described. In researching regional memories of disasters, we

[3] Waters-Bayer, A., Local innovation, and participatory innovation development: The concepts and their applications in livestock-related research and development, in participatory innovation and research: Lessons for livestock development, ESAP Proceedings 2005.

[4] Hoeggel, F. U. and Mbeyale, G., *Impact assessment of pastoralist field schools in Ethiopia and Uganda*, FAO, SADC, 2014.

utilized the social memory of oral historians and time experts to investigate failure, reorganization, and adaptive diversity. In discussing socio-political institutions as a repository for social memory in passing knowledge from one generation to another, the Borana *gada* timeline was used to construct a traditional event history calendar (EHC) to recount past disasters that contributed to the collapse of pastoralism (Chap. 3). The traditional EHC is used to recall past disasters associated with megadroughts, famines, epidemics, wars, locust plagues, and periods of excessive wetness, adversely affecting pastoralism. The cyclical historical events serve as a yardstick to recall the past and preserve the present. The mechanisms for building social memory surrounding historical events are connected to pastoralists' mobile land-use practices. The crises associated with certain disasters transformed human adaptations over the years, while post-disaster periods were times of reconstituting socio-political institutions and resource allocations to repair the damage that had been suffered.

The work additionally presents the political and climate history of the southern Ethiopian–northern Kenyan frontier and the resulting pastoral adaptations (Chap. 4). Practically, the analysis uses geographical space to characterize access across international borders with their restrictions to trans-frontier grazing and watering by pastoralist herds. The causes were political contentions between Ethiopia and the British and later the British and Italian administrations to maintain pre-colonial grazing movements. The control of pastoralist movements ignored the compounding effects of climatic variability. The lengthy historical analysis shows that wet years were interspersed with droughts, famine, and periods of locusts and epidemics, all of which influenced the movements of pastoralist herds. Rainfall patterns in northern Kenya varied from patchy to totally lacking to extremely heavy. During the five decades, rainfall deviations from the long-term mean show droughts occurring 56.8 percent of the time, relative to above-average rainfall that occurred 43.2 percent of the time. Over the period (1908–1962), livestock populations and rangeland production fluctuated between scarcity and surpluses. By the time of Kenya's independence in 1963, a secessionist war triggered a collapse of pastoralism, forcing the reorganization of diverse adaptive strategies.

Reorganization and Adaptive Diversities

The wars marked a significant break from the past regarding land-use change (Chap. 5). Land-use change processes were linked through various pathways reflecting significant economic reorganization. The Obbu farmers invested time in crop cultivation and managing private range enclosures. This land tenure transformation into private farms and range enclosures was motivated by expectations of future economic benefits. The extent of enclosed rangelands had expanded by the 2000s. Accordingly, this study suggests that two factors may be attributed to a functional, sustainable, and resilient agropastoral ecosystem. First, periodic food production at least once every four years was a valuable substitution for agropastoral household livelihoods. Second, farming systems included the accumulation of fodder reserves rented out or saved as feed reserves for stock during periods of stress. However, land fragmentations associated with crop cultivation and range enclosures affected individual drought management strategies.

We examined how changes in land use influenced drought management strategies and decision-making by individual and cooperative herding units (Chap. 6). We discussed histories of drought management strategies by herders of different wealth ranks, focusing on their adaptive decisions associated with herd movements. Seasonality and rainfall variability triggered these decisions, including water distribution, pasture, and political conflicts. Individual and cooperative herding unit's mobility patterns changed from one drought stress period to another. Their decisions were influenced by access to labor, which is necessary for herd splitting into mobile and home herds. Long-distance herd movements at the peak of drought caused more stress for cattle since they spent more energy walking—exacerbating their already drought-weakened conditions.

Drought management strategies varied among large and small herd owners. Large herd owners often adopted regional adaptive strategies, while small herd owners practiced local and intensive drought management strategies. By the 1990s, long-distance herd movements across the international frontier had become rarer. By the 2000s, the grazing lands associated with milk herds had disappeared (because of expanding private range enclosures), while the rangelands used by mobile dry packs had shrunk by 30 percent.

The community's adaptive responses to droughts were three-fold. First, the emergence of cooperative herding meant that households could pool their labor resources. Second, the disappearance of home-grazing lands forced the community to return to mobile herd management strategies. Third, the distribution of settlements in areas where the community was displaced by the war five decades earlier enabled them to create a hybrid management system that combined indigenous grazing associations (*dedha*) and formal pasture conservation (*mazingira* in Kiswahili). The purpose is to save stock during drought years.

Drought frequencies and management significantly impacted the cattle population and shift in household wealth (Chap. 7). The effects of decadal droughts on trajectories of cattle demography and household wealth were monitored for 29 years. We framed the analyses regarding the resilience of household cattle demographics and wealth dynamics. We examined pre- and post-drought periods and land-use intensification across agroclimatic zones among the eight study sites and at local scales. We investigated the impacts of decadal droughts on herd structure and household wealth dynamics between periods of collapse and recovery. Rates of herd recovery depended on the time intervals between droughts. Herd dynamics varied across agroclimatic zones and localities. Survival trends were more significant among reproductive females and immatures, implying that the cattle population persisted despite the herd population collapsing. There were positive correlations between annual rainfall levels and cattle population—when rainfall was below average, herd population collapsed.

Using household wealth life histories, we tracked cattle wealth dynamics in terms of cattle per capita and shifts in wealth rank in response to decadal droughts. The data shows that over the monitoring period, wealth recovery remained incomplete, particularly after the droughts of 1991/92 and 2011. No households remained in their pre-drought wealth ranks in the six life histories we monitored over 12 years. The most volatile wealth ranks were those of the wealthy, the rich, and medium families, which disappeared during droughts, with some even becoming stockless. A collapse in cattle population and household wealth affected social capital networks.

Collapse and Transformation of Social Capital Networks

The work further discusses how the Obbu Borana showed flexibility in transforming social capital initiatives to spread risks and support the stockless households (Chap. 8). In the past, sharing was between different social network zones in northern Kenya (home) and across the border in Ethiopia (hosts)—this required coordination between sub-clan members in both regions. First, the help for the needy was considered and approved in stages. The participants observed common customs and regulations used to enforce decisions. Second, widespread political representation at local levels allowed everyone to benefit from sub-clan assistance. Third, social stock-sharing criteria were based on wealth levels that divided the society into givers, receivers, and non-participants. The givers were the rich and those of medium wealth. The beneficiaries were stockless households. Negotiations were lengthy and time-consuming. Strict adherence to rules and regulations made the system highly democratic but lacking flexibility. Using the narratives of clan representatives, we conclude that the clan social capital network had become dysfunctional.

The adaptive response was establishing a new broad-based neighborhood social capital network system based on a diversified economy. The transformation was done progressively. First, clans formed alliances to pool communal resources. The second step was to create a neighborhood social capital network coordinated by organizing committees[5] registering those requiring societal assistance. Everyone is a contributor to broad-based social capital networks. Both the rich and the poor—may benefit from fundraising activities. Increased participation in non-pastoral economies and the use of cash payments have transformed the concept of poverty. Individuals have changed social obligations and entitlements in the new neighborhood social capital networks, making the system highly adaptive and resilient.

We further discussed the resilience of women-operated neighborhood household food security networks (Chap. 9). The characteristic features of individual household food security networks included sharing relations with relief organizations. The links between rural and peri-urban household networks allowed women to exploit diversities of resources. Using income from market supplies of dairy and grain, women developed networks of debts at the food shops and kiosks that sustained food security. By combining informal and formal networks, women in rural and peri-urban environments increased their access to community-wide resources. They raised their access to extra resources, mainly using public fundraising and linkages to the retail trade and food shops. Quantities of food exchange and their monetary values varied among households of different wealth ranks. An exciting aspect of the findings is the connections between food security and household wealth. Regarding grain, medium wealth households solicited more than rich or poor households. The rich have sufficient

[5] The committees comprising both genders were elected by the neighborhoods. They comprised chairperson, secretary, treasurer, and members. They oversaw public fundraising and inviting guests of 'honors'—often moneyed individuals and people in leadership.

cash to offset deficits, compared to the poor, who lack that capacity. Accordingly, as a rule of thumb, the poor keep within their means and solicit amounts that they would be able to give back in the future, should it become necessary. From the analyses, there are emerging issues that might be given more attention for achieving progressive strategies for mitigating droughts in the future.

Progressive Opportunities for Mitigating Droughts

Progressive opportunities comprised various innovations that herders and farmers used to mitigate droughts.[6] Innovations are locally developed methods—often created through trials. Consequently, in participatory innovation, communities conduct experiments, adopt what worked, and disregard those that failed. Innovations can promote resilience through livelihood-specific actions.[7] From the findings in Chaps. 3–9, actionable innovative strategies for mitigating drought impacts include social memory relying on the *gada* timeline (Chap. 3) and mobility in response to governance and climate history (Chap. 4). Others include the establishment of agropastoralism, growing fodder reserves (hereafter fodder banks), and changes in resource tenure (Chap. 5). Others include reorganizing traditional institutions for managing communal grazing lands and the new *mazingira* conservation systems (Chap. 6). Additional actionable management strategies are linked with the re-establishment of mobile herding through corporate and hired herding labor,[8] while others are linked to saving stock losses during drought years (Chap. 7). Operations of neighborhood social capital networks and evaluation of their efficacy for mitigating droughts (Chap. 8) and women's social capital networks' impact on neighborhood household food security (Chap. 9). Finally, the herders and farmers require training with practical experiences to improve drought survival strategies. We will briefly discuss these actionable innovations for future drought mitigations in turn.

[6] Pickering, T., *Drought, and Samburu pastoralism: A collaborative effort to examine the causes, adaptation process, and grazing innovations in Northern Kenya*, PhD Dissertation, Colorado State University, Fort Collins, Colorado.

[7] McPeak, J. G. and Little, P.D., 'Applying the concept of resilience to pastoralist household data,' *Pastoralism: Research, Policy, and Practice* (2017) 7:14, https://doi.org/10.1186/s13570-017-6082-4.

[8] Ensminger, J., 'Reputations, trusts, and principal agent problem,' in Cook, Karren, S. (ed.) *Trust in society* (New York, Russel Sage Foundation, 2001).

Social Memory as a Tool for Predicting Droughts

We have used social memory and proxy environmental indicators to reconstruct EHC. The calendar constructed using the Borana *gada* timeline provided how the Borana pastoralists predicted the return cycles of droughts, famine, wars, locust plagues, and epidemics. In this section, we combined the historical EHC and the meteorological data from the colonial archive to complete the calendar to include the periods covered by the case studies reported in this work (Chaps. 4–7). We conclude from the earlier and present periods that the years with low rainfall correspond with periods of high stock mortalities, while above-average rain is followed by herd recovery. The data suggests that while periods of cattle losses during excessive wet weather occur every 10–15 years, droughts and famine have returned more frequently. The drought cycles and their returns according to the *gada* and *maqbas* cycles, including those of earlier periods and the twentieth and the first quarter of the twenty-first century, are summarized in Table 10.1.

The EHC predicts 5–8 years, 50 years, and 100 cycles of droughts connecting the past with the present. The 100-year cycle that returned during the 1991/92 famine was a return when blue flies, like the periods of the Rinderpest pandemic, infected cattle, causing mass mortality. While there was no evidence of rinderpest during the latter famine, the blue flies reported earlier in oral sources aggravated the famine conditions. The *daacci* of cattle pandemic returned during the *gada* of Liban Kusse (1913–1921) when cattle died of contagious bovine pleuro-pneumonia (CBPP) (see Chap. 4). The 100-year drought cycle of the *Fullasa* of Kura Jarso (2016–2024) is a return of three-year droughts during the *gada* of Boru Galma (1904–1912) of the same *maqbas* (Table 10.1). The droughts during the *gada* of Liban Jaldesa (1888–1896), Bulle Dabasa (1929–1936) and Goba Bulle (1968–1976), and Guyo Goba (2008–2016, see Tables 3.1 and 10.1) returned to the same *gogessa*. Further, the *Moggisa* of Guyo Boru (1944–1952) that alternated between recovery and drought returned during the *gada* of Jaldesa Guyo Dabasa (1848–1856) (see Table 3.1). As mentioned in Chap. 3, the Borana oral historians report cyclic returns of droughts associated with a solar eclipse. The *Libasa* of Bulle Dabasa (1929–1937) is related to severe droughts that killed livestock and people (see Chap. 4) after a solar eclipse. The next solar eclipse occurred during the *gada* of Goba Bulle (1968–1976), coinciding

Table 10.1 Drought cycles recorded by meteorological data against gada time for the twentieth and the first quarter of the twenty-first century

Gada timeline	*Years*	*Maqbas cycle*	*Events*
Adi Doyo Jillo	1888–1904	*Makula*	Epidemic, famine
Boru Galma Doyo	1904–1912	*Fullasa*	Epidemic, droughts[a]
Liban Kuse	1912–1920	*Mardida*	Wet, droughts
Arero Gedo	1920–1928	*Darara*	Regional drought
Bulle Dabassa	1928–1936	*Sabbaqa*	Drought, wet
Aga Adi Doyo	1936–1944	*Sabbaqa*	Famine
Guyo Boru	1944–1952	*Moggisa*	Recovery[b]
Mada Galma	1952–1960	*Makula*	Drought, wet
Jaldesa Liban	1960–1968	*Fullasa*	Wet, famine[c]
Goba Bulle	1968–1976	*Mardida*	Famine[d]
Jillo Aga Adi	1976–1984	*Darara*	Drought, wet
Boru Guyo	1984–1992	*Libasa*	Famine[e]
Boru Mad'a	1992–2000	*Sabbaqa*	Famine[f]
Liban Jaldesa	2000–2008	*Moggisa*	Recovery
Guyo Goba	2008–2016	*Fullasa*	Severe drought
Kura Jarso	2016–2024	*Fullasa*	Famine[g]

[a]The 100-year cycle of this drought returned in 2021
[b]This was referred to as the drought of hides and skins,
[c]International famine relief for the first-time distributed fish to the pastoralists
[d]This coincided with the Sahelian disaster of the early 1970s
[e]The return cycle of a severe drought when massive relief saved human lives was during the 1983/drought
[f]This was the 100-year-return cycle of mass cattle caused by rinderpest. During the later famine, the blue flies that fed on the carcasses of animals during the earlier drought were evident
[g]The famine conditions were a return of 100-year cycle of the disasters of 1891, in terms of stock mortalities

with the infamous Sahelian disaster in 1973.[9] The *sabbaqa* of Walle Wacuu (1713–1721) returned with Boru Mad'a (1992–2000) after 280 years. These were both periods of famine (see Tables 3.1 and 10.1). While the returns of the drought/famine year cycles appear predictable, the grazing lands in the Obbu region have radically changed. The changes triggered by warfare in the 1960s fundamentally altered land tenure by adopting agropastoralism.

[9] Sinclair, A. R. E., and J. M. Fryxell. "The Sahel of Africa: ecology of a disaster." *Canadian Journal of Zoology* 63.5 (1985): 987–994.

Changes in Resource Tenure and Use of Fodder Banks

Triggers that fragmented grazing lands in the Obbu region are changes in resource tenure prompted by private farms and range enclosures. The purpose is to preserve fodder for dry year use. This section will discuss how this locally innovated technology evolved to manage access to hay at local levels. We will briefly explain the phases of land-use change. Phase 1 coincided with the pre-secessionist war when land use was dominated by communal grazing (the 1960s–1970) when access to the commons was unrestricted. Phase 2 (1970–1980) corresponds with a pre-agropastoral land use whereby households continue to graze extensively, but a fence in small plots near their homesteads that are used for growing crops that double as a place for animals to graze if the crops fail or after the harvest. During Phase 3, there is a clear differentiation of land use into crop cultivation and preservation of pasture. Phase 3 occurred when land use was transformed (1980–1990)—farming combined with enclosures fenced off 46 km^2 (about 28.58 mile^2) strip along the frontier. Due to the growing market demands for fodder, the focus in Phase 4 is on fodder banking (1990–2000), which overlaps with the expansion of private range enclosures. During the latter period, more than 660 km^2 of grazing lands had been fragmented by range enclosures. Between 2000 and 2016, crop cultivation declined in most areas due to the unreliability of rainfall, while the expansion of enclosures had increased—driven by future monetary benefits. Increased investments in fodder banks are the safest solution for solving forage scarcity.

Building Fodder Banks

Building fodder banks is critical for agropastoral households that often experience scarcities during the dry season. Considering the trends of fodder commercialization, this calls for innovative methods for large-scale fodder production. Haymaking is one such innovation.[10] On finer scales, hundreds of such enterprises exist across the Obbu region of northern Kenya. The standing hay in farms and enclosures during heavy rainfall

[10] Waters-Bayer, A., Local innovation, and participatory innovation development: The concepts and their applications in livestock-related research and development, in participatory innovation and research: Lessons for livestock development, ESAP Proceedings 2005.

might encourage the communities to build fodder banks.[11] Still, traditional hay needs to be higher in nutritional value, requiring innovative technologies for harvesting grass at stages of plant growth when forage quality is high. For enclosure owners, fodder markets provide an essential source of income. This means that there are prospects to justify the development of fodder banks. We can now suggest how fodder banks may be integrated with managing rangelands to mitigate forage scarcities during stress periods.

Application of the Mazingira Conservation and Management System

Grazing lands are managed by grazing associations managed by elder councils (*jaarsa dedha*). The association coordinates grazing across patchy landscapes, allocating grazing resources between wet and dry seasons. In recent decades, governance systems by this indigenous resource management have weakened as the state reorganizes the community's governance.[12] Nonetheless, by combining informal and formal institutions, the Obbu Borana have developed a hybrid system for pasture conservation. Briefly, the system works as follows. Following land fragmentations along the frontier, the primary grazing is concentrated in the arid plains of Golbo (see Fig. 2.3). During recent decades, after the development of boreholes and pans, access to water has improved. This encouraged the expansion of permanent settlements in the zone. It is around these areas that the systems of *mazingira* for conserving pasture were conducted by individual settlements. However, research has not evaluated the system's efficiency; the issues were discussed during workshops organized by this author. The strategy is that during the growing season, the grazing in selected landscapes was preserved for the dry season grazing. Rules on access rights and the consequences of violating the regulations were vigorously enforced through heavy fines.

There is still a challenge in using the *mazingira* system. The rainfall being highly erratic, rains might fail in the protected landscapes while it rains elsewhere. Under the circumstances, the settlements would

[11] Beyene, S. and Gudina, D., 'Reviving a traditional pasture management system in Fentale, East central Ethiopia,' *Journal of Ecological Anthropology* 13 (1) (2009): 57–72.

[12] Bassi, M., 'The complexity of a pastoral African polity: An introduction to the council organization of the Borana,' *Journal of Ethiopian Studies* 32 (2) (1999): 15–33.

negotiate with those neighborhoods that received rainfall. The solution is for different settlements to jointly preserve grazing in areas supplied with boreholes. Keeping feeding around the borehole rangelands during the wet season is the strategy used by the *dedha* associations. By combining the management of *mazingira* and the development of the fodder banks, farmers could bale hay in sufficient quantities to save stock during drought years. Still, due to the scarcity of animal feeds, the Obbu region would remain a net importer for some time. During the 2021 and 2022 droughts, wealthy individuals imported hay from other parts of Kenya and sold it locally. From personal observations, costs for maintaining a stock of 40 heads on the primary hay feed are estimated at USD 1000 per month.[13] The cost is beyond the financial means of most households. An increase in the market price of animal feed will be offset if those owning range enclosures get involved by bulking fodder to mitigate stock losses for drought years. The shrinking grazing commons associated with fragmentation by crop cultivation and range enclosures implied that traditional rangeland management needed to maintain its historical flexibility. Regardless of the changes, the overarching climate events associated with famine and drought cycles are more severe on the stock than usual. This will result in more significant stock losses with direct consequences on the functioning of social capital networks.

Evaluations of Social Capital Networks

During the pandemic of the 1890s, when the traditional clan social capital network of *buusa gonoofa* was broken, the Borana resuscitated the system following recovery (see Chap. 3). Unlike the late nineteenth century, the current situation represents a collapse and transformation of the clan social capital networks (Chap. 8). Consequently, the clans joined to establish a broad-based network. The new system is organized by neighborhoods, officially mandated by government officials (through chiefs' offices) to raise funds from the public. Meanwhile, individuals using various modern communication technologies (telephones, WhatsApp groups, e-mails, and Facebook) would broadcast the date by approaching neighborhood members for support. Those not in the locality use the Safaricom money

[13] This is the cost of providing hay and other animal feed to one family herd in 2022.

transfer system of '*Mpesa*'[14] to transfer amounts they pledged. Mapping the neighborhood networks and understanding types of support would help know the extent to which the system may be used to mitigate drought impacts in the future.

Among the new broad-based social capital system beneficiaries are women, whose traditional social capital network membership is open to neighbors, friends, and relatives for sharing food items and services (Chap. 9). Through formal organizations, women's groups have successfully attracted funding from external sources and public funds, which they combined with traditional *maaro* networks to improve household food security. Through networks of debt-client systems with food shops and food kiosks, women improved their access to food. The disadvantage is for those who lost stock during droughts. Their failure to repay debts will adversely affect business, forcing them to be closed or cause a stock decline. With this background, it might be appropriate to ask about the prospects for drought management by the Obbu Borana.

Prospects and the Future

In this final section, we will examine prospects for the future. Using multiple pathways of livelihoods, the Obbu Borana households have various choices. The region is located along the international highway linking Kenya to Ethiopia. Improved communication is rapidly opening the region for business. For the first time, the main town of Sololo-Ramata is served by a bank, presenting evidence of increased money circulation. More business enterprises and new job opportunities were created following the implementation of devolution after the promulgation of the 2010 Kenyan constitution. The region now receives more significant national financial allocations than in the past, which has a positive trickle-down effect. Besides, the welfare of individual households is improving. Through motorbike transport (*Booda booda*), youths have gained self-employment to ease transportation in rural areas. The availability of wide-ranging consumer goods and well-connected mobile phone networks to the rest of the

[14] Asfaw, Solomon, Benjamin Davis, Josh Dewbre, Sudhanshu Handa, and Paul Winters, 'Cash transfer programme, productive activities and labour supply: evidence from a randomised experiment in Kenya,' *The Journal of Development Studies* 50, no. 8 (2014): 1172–1196.

world has made it possible for the Obbu households to participate in the Safaricom money transfer (*Mpesa*).

Still, these new opportunities will be improved if it's translated into field schools and training organized for the herders and the farmers. Utilizing various innovative strategies, they developed methods to mitigate droughts; their efforts would be more worthwhile if they were trained to improve their strategy. The Obbu region will be used as a pilot to develop informal training to integrate pastoralists and farmers to promote community adaptive innovations to manage climate variability. A prospective action being proposed is to organize a pastoralist education center by providing environmental and social-economic context by developing a curriculum for lifelong education. It is anticipated that through training, the community will be helped to improve their knowledge by focusing on new technologies for future mitigation of droughts. Through regular seminars, workshops, and demonstrations of innovative technologies, discussions with the community will be improved for developing a vision for the future. The skills gained from regular training will snowball to the rest of the community to regenerate knowledge for better survival during future droughts. The educational curriculum offered at the training center will be designed with specific groups of pastoralists and agropastoralists in mind, utilizing their unique cultural and economic opportunities to facilitate their future adaptive strategies. The center will develop learning tools and methods to promote innovations and learning outcomes by serving as a herder and farmer indigenous knowledge depository. Connected to the learning and educational center shall be a community library—a valuable resource for supporting field demonstration and offering additional opportunities for tertiary institutions conducting research in the region to foster innovative indigenous adaptive strategies.

Bibliography

Abel, N., Cumming, D. H. M. and Anderies, J. M., 'Collapse and reorganization in social-ecological systems: Questions, some ideas, and policy implications,' *Ecology and Society* 11.1 (2006): 17 [online] http://www.ecologyandsociety.org/vol11/iss1/art 17/.

Adebayo, R. M., Adeoye, A. S., Omisore, O. A. and Alabi, A. O., 'Indigenous institutions for collective action in fostering peace for sustainable land management among pastoralists and crop farmers in Ogun state, Nigeria,' *Nigerian Journal of Rural Sociology* 16 (3) (2016): 56–63.

Adger, W. N., 'Social and ecological resilience: Are they related?' *Progress in Human Geography* 24 (3) (2000): 347–364

Adger, W. Neil, Suraje Dessai, Marisa Goulden, Mike Hulme, Irene Lorenzoni, Donald R. Nelson, Lars Otto Naess, Johanna Wolf, and Anita Wreford. 'Are there social limits to adaptation to climate change? *Climatic Change* 93 (3) (2009): 335–354.

Ado, A. M., Savadogo, P., Abdoul-Azize, H. T., 'Livelihood strategies and household resilience to food insecurity: Insight from a farming community in Aguie district of Niger,' *Agriculture and Human Values* 36 (2019): 747–761.

Agarwal, S. K., 2013, 'Emerging technological intervention models with scalable solutions for adaptions to climate change and livelihood gains in Indian Himalayan region: Case studies on action research at the grassroots level,' in Nantiyal, S. et al. (Eds.), *Knowledge systems of societies for adaptation and mitigation* (Berlin: Springer-Verlag).

Agrawal, A. and Perrin, N., *Climate adaptation, local institutions and rural livelihoods*, IFRI Working Paper # W08I-6.

G. Oba, *Pastoralist Resilience to Environmental Collapse in East Africa since 1500*, https://doi.org/10.1007/978-3-031-48291-5

Ahmed H., Tessema, Z., Tolera, A. and Korecha, D., 'Inter-connection between El-Niño Southern Oscillation induced rainfall variability, livestock population dynamics and pastoralists adaptive strategies in Eastern Ethiopia,' *Journal of Environment and Earth Sciences* 7 (3) (2017): 11–24.

Akyyeampong, E., Bates, R. H., Nunn, N. and Robinson, J. A. 'Introduction: Africa-the historical roots of its under development,' in Akyeampong, E., Bates, R. H., Nunn, N. and Robinson, J. A. (Eds.) *Africa's development in historical perspective* (New York: Cambridge University Press, 2014).

Aldrich, D. P., "Fixing recovery: Social capital in post-crisis resilience." *Journal of Homeland Security, Forthcoming* (2010a).

Alemayehu, K. and Fantahun, J., 'The effect of climate change on ruminant livestock population dynamics in Ethiopia,' *Livestock Research for Rural Development* 24 (10) (2012): 1–10.

Aldrich, D. P., *Fixing recovery: Social capital in post-crisis resilience*. Purdue University-Purdue-Pubs 5-2010b, Department of political science.

Aldrich, D.P., Meyer, M. A., 'Social capital and community resilience,' *American Behavioral Scientist* 59 (2) (2015): 254–269.

Ali, S. M., *Huddur and the history of southern Somalia* (Cairo: Nahda Bookshop Publisher, 2005.

Amare, A. and Simane, B., 'Does adaptation to climate change and variability provide household food security? Evidence from Muger sub-basin of the upper Blue-Nile, Ethiopia,' *Ecological Processes* (2018): 7: 13, https://doi.org/10.1186/S13717-018-0124x.

Anderson, David M., *Resilience and Collapse in African Savannahs: Causes and consequences of environmental change in east Africa* (London: Routledge, 2018);

Anderson, D. M., 'The beginning of time? Evidence for catastrophic drought in Baringo in the early nineteenth century,' in Anderson, D. M. and Bollig, M. (Eds.) *Resilience and collapse in African savannahs: Causes and consequences of environmental change in East Africa* (London: Routledge, 2017).

Anderson, D. M., Elliott, H., Kochore, H. H. and Lochery, E., 'Camel herders, middle women, and urban milk bars: The commodification of camel milk in Kenya,' *Journal of Eastern African Studies* 6 (3) (2012): 383–404.

Andrezjewski, B. W., 'Images of drought in Somali literature,' in Lewis, I. M. (Ed.) *The Somali drought* (London: IAI, 1975, 15).

Angassa, A. and Oba, G., 'Relating long-term rainfall variability and cattle population dynamics in communal rangelands and a government ranch in southern Ethiopia.'

Angassa, A. and Oba, G., 'Herder perceptions on impacts of range enclosures, crop farming, fire ban and bush encroachment on the rangelands of Borana, Southern Ethiopia,' *Human Ecology* 36.2 (2008): 201–215.

Ansah, G. K., Gardebroek, C. and Ihle, R., 'Resilience and household food security: A review of concepts, methodological approaches and empirical evidence,' *Food Security* 11 (2019): 187–1203.

Arjjumend, H., 'Review of pastoralists' resilience and adaptation to climate change: Can technology help pastoralists mitigate the risks? *International Journal of Bio-resource and stress Management* 9 (1) (2018): 118–128.

Asfaw, Solomon, Benjamin Davis, Josh Dewbre, Sudhanshu Handa, and Paul Winters, 'Cash transfer programme, productive activities and labour supply: evidence from a randomised experiment in Kenya,' *The Journal of Development Studies* 50, no. 8 (2014): 1172–1196.

Asiwaju, A. I. 'Borders and borderlands as linchpins for regional integration in Africa: Lessons of the European experience.' *Africa Development/Afrique Et Développement* (1992): 45–63.

Ayana, E. K., Ceccato, P., Fisher, J. R. B. and DeFries, R., 'Examining the relationship between environmental factors in pastoral areas of East Africa,' *Science of the Total Environment* 557–558 (2016): 601–611.

Ba, A., Lesnoff, M., Poccard-Chapuis, R., Moulin, C-H., 'Demographic dynamics and off-take of cattle herds in southern Mali,' *Tropical Animal Health Production* 43 (2011): 1101–1109.

Badejo, A. F., Majekodunmi, A. O., Kingsley, P., Smith, J. and Welburn, S. C., 'The impact of self-help groups in pastoral women's empowerment and agency: A study in Nigeria,' *Pastoralism, Research, Policy & Practice* 7 (2017): 28, https://doi.org/10.1186/s13570-017-0101-5.

Badzaban, F., Rezaei-Moghaddanam, K. and Fatemi, M., 'Analysing the individual entrepreneurial resilience of rural women,' *Journal of Global Entrepreneurship Research*, https://doi.org/10.1007/s40497-021-00290-1.

Baird, Timothy D., and Clark L. Gray. "Livelihood diversification and shifting social networks of exchange: a social network transition?" *World Development* 60 (2014): 14–30.

Baival, B., *Community-based rangeland management and social-ecological resilience of rural Mongolian communities.* Doctor of Philosophy, Department of Forest and Rangeland Stewardship, Colorado State University (2012).

Barth, F., *Models of social organization.* Occasional Paper, Royal anthropology Institute 23 London: Royal Anthropological Society.

Barton, D., Meadows, N. and Morton, J., *Drought losses, pastoral saving, and banking: A review*, DFID Advisory and Support Services Commission Project 2W0027 NR Project L0114/2001.

Basset, J. J., 'The political ecology of peasant-herder conflicts in the northern ivory coast,' *Annals of the Association of American Geographers* 78 (3) (1988): 453–472.

Bassi, Marco. "The System of Cattle Redistribution among the Boraana Obbuu and its Implications for Development Planning.' In Baxter P.T. W. and Hogg,

R. (eds). *Property, poverty, and people: changing rights in property and problems of pastoral development* (Department of Social Anthropology, University of Manchester, 1990).

Bassi, M., 'The complexity of a pastoral African polity: An introduction to the council organization of the Borana,' *Journal of Ethiopian Studies* 32 (2) (1999): 15–33.

Bates, D. G., 'The role of the state in peasant-nomad mutualism,' *Anthropological Quarterly* 44 (3) (1971): 109–131.

Bauer, K. M., *High frontiers: Dolbo and the changing world of Himalayan pastoralists* (Colombia University Press, 2004).

Baxter, P.T.W. 'Ethiopia's unacknowledged problem: the Oromo,' *African Affairs* 77, no. 308 (1978): 283–296.

Beaman, Anne W. 'Women's participation in pastoral economy: Income maximization among the Rendille.' *Nomadic Peoples* (1983): 20–25.

Behnke, Roy H., and Carol Kerven. *Redesigning for risk: tracking and buffering environmental variability in Africa's rangelands.* Vol. 1. (London: Overseas Development Institute, 1994).

Behnke, Roy H., Ian Scoones, and Carol Kerven, eds. *Range ecology at disequilibrium: new models of natural variability and pastoral adaptation in African Savannas.* (London: Overseas Development Institute, 1993).

Belli, Robert F., William L. Shay, and Frank P. Stafford. 'Event history calendars and question list surveys: A direct comparison of interviewing methods,' *Public Opinion Quarterly* 65.1 (2001): 45–74.

Belay, K., Beyene, F. and Manig, W., 2005. 'Coping with drought among pastoral and agro-pastoral communities in Eastern Ethiopia,' *Journal of Rural Development* 28: 185–210.

Belli, Robert F. 'The structure of autobiographical memory and the event history calendar: Potential improvements in the quality of retrospective reports in surveys.' *Memory* 6.4 (1998): 383–406.

Beinart, W. and Coates, P., *Environment, and history: The taming of nature in the USA and South Africa* (London: Routledge, 1995).

Béné, Christophe, Derek Headey, Lawrence Haddad, Klaus von Grebmer. 'Resilience a useful concept in the context of food security and nutrition programmes? Some conceptual and practical considerations.' *Food Security* 8, no. 1 (2016): 123–138.

Bessens, I., Verschuren, D., Russell, J. M., Hus, J., Mees, F. and Cumming, B. F., 'Paleolimnological evidence for widespread late 18th century drought across equatorial East Africa,' *Paleoclimatology* 259 (2008): 107–120.

Berhanu, W., *Recurrent shocks, poverty traps and the degradation of the social capital base of pastoralism: A case study from southern Ethiopia*, Center of Evaluation for Global Action Agriculture for Development, Paper No. AFD-0916, issued in July 2009.

Berzborn, S. and Schnegg, M. 'Vulnerability, social networks, and resilience: Rural livelihoods in the Richtersveld, South Africa, Research Gate,' https://www.researchgate.net/publication/309616695.

Beyene, K., 'Exploring incentives for rangelands enclosures among pastoral and agropastoral households in eastern Ethiopia,' *Global Environmental Change* 19 (2009): 484–502.

Beyene, S. and Gudina, D., 'Reviving a traditional pasture management system in Fentale, East central Ethiopia,' *Journal of Ecological Anthropology* 13 (1) (2009): 57–72.

Beyene, F., 'Interclan cooperation in a risky pastoral ecology: some lessons from Eastern Ethiopia,' *Human Ecology* 38 (1991): 555–565.

Blackburn, Morgan, Alejandro Guzman, Jeff Lieberman, and Anne Sprinkel. "Assessing the role of social capital in agro-pastoral resilience in the Sahel: A systems Perspective." *coordination with Mercy Corps and the Elliott School of International Affairs, George Washington University* (2014).

Bollig, Michael, and David M. Anderson, eds. *Resilience and Collapse in African Savannahs: Causes and consequences of environmental change in east Africa* (London, Routledge, 2018).

Bollig, Michael. 'Adaptive cycles in the savannah: pastoral specialization and diversification in northern Kenya.' *Journal of Eastern African Studies* 10.1 (2016): 21–44.

Bolling, M. and Lang, H. 1999 'Demographic growth and resource exploitation in two pastoral communities' (*Nomadic Peoples* 3, 16–34).

Bollig, M. 'Moral economy and self-interest: kinship, friendship, and exchange among the Pokot (N.W. Kenya),' In Schweizer, T. and White, D.R. (eds.) *Kinship. Networks, and exchange* (Cambridge University Press, 1998).

Boone, R. B. and Wang, G., 'Cattle dynamics in African grazing systems under variable climates,' *Journal of Arid Environments* 70 (2007): 495–513.

Bottrell, Dorothy, 'Dealing with disadvantage: resilience and social capital of young people's networks,' *Youth and Society* 40 (4) (2009a): 476–501.

Bottrell, Dorothy. 'Understanding 'marginal' perspectives: Towards a social theory of resilience.' *Qualitative social work* 8.3 (2009b): 321–339.

Branch, D., 'Violence, decolonization and the cold war in Kenya's North-Eastern Province, 1963–1978' *Journal of Eastern African Studies* 8 (4) (2014): 642–657.

Breshears, David D., López-Hoffman, L. and Graumlich, L. J., 'When ecosystem services crash: preparing for big, fast, patchy climate change.' *Ambio* 40.3 (2011): 256–263.

Brown, M. *Where Giants trod: The saga of Kenya's desert lake* (London: Quiller Press, 1989).

Burian, A., Karaya, R., Wernersson, J. E. V., Egberth, M., Lokorwa, B. and Nyberg, G., 'A community-based evaluation of population growth and agro-pastoralist resilience in Sub-Saharan drylands,' *Environmental Science and Policy* 92 (2019): 323–330.

Brunson, M. W., 'The elusive promise of social-ecological approaches to rangeland management,' *Rangeland Ecological Management* 66 (2012): 632–637.

BurnSilver, S., 'Presenting wealth in a changing pastoral economy: A comparison of traditional and new metrics in Maasailand, Kenya,' *Nomadic Peoples* 20 (2016): 8–34.

Bushyby, K. and Sites, E., *Resilience and risk in pastoralist areas: Recent trends in diversified and alternative livelihoods, Karamoja, Uganda* in USAID, 2016.

Butzer, K. W. 'Collapse, environment, and society,' *Proceedings of the National Academy of Sciences* 109.10 (2012): 3632–3639.

Cabell, J. F. and Oelofse, M., 'An indicator framework for assessing agroecosystem resilience,' *Ecology & Society* 17 (1) (2012): 18, https://doi.org/10.5751/ES-04666-170118.

Ca'ceres, D. M. and Gras, C., 'A tipping point for agricultural expansion? Technological changes and capital accumulation in Argentina's rural sector,' *Journal of Agrarian Change* 20 (2020): 79–93.

Campbell, D. J., Lusch, D. P., Smuker, T.A. and Wangui, E.E. 2005. 'Multiple methods in the study of driving forces of land use and land cover change: A case study of S.E. Kajiado, District, Kenya,' *Human Ecology* 33: 763–794.

Campbell, D.J., Response to drought among farmers and herders in southern Kajiado District Kenya: A comparison of 1972–1976, *Human Ecology* 27 (3) (1999): 377–415.

Carpenter, S. R., Brock, W. A. and Ludwig, D., 'Collapse, learning, and renewal,' in Gunderson, Lance H., Craig Reece Allen, and Crawford S. Holling, eds. *Foundations of ecological resilience*. Island Press, 2012.

Cassidy, L. and Barnes, G. D., 'Understanding household connectivity and resilience in marginal rural communities through social network analysis in the village of Habu, Botswana,' *Ecology & Society* 17 (4) (2021): 11, https://doi.org/10.5751/ES-04963-170411.

Caviedes, C. N., *El Niño in history. Storming through ages* (Florida, FL: University Press of Florida, 2001).

Chang C. and Koster, H. A., *Pastoralists at the periphery: Herders in a capitalist World* (Tucson: The University of Arizona Press, 1994).

Cleaver, Frances. "The inequality of social capital and the reproduction of chronic poverty." *World Development* 33.6 (2005): 893–906.

Cumming, G. S. and Peterson, G. D., 'Unifying research on social-ecological resilience and collapses,' *Trends in Ecology and Evolution* 32 (9) (2017): 695–713.

Collins, S. L., Carpenter, S. R., Swington, S. M., Orenstein, D. E. et al., 'An integrated conceptual framework for long-term social-ecological research,' *Frontiers of Ecological and Environment* 9.6 (2011): 351–357.

Conway, D., 'Extreme rainfall events and lake level changes in East Africa: Recent events and historical precedents,' in Odada, E. O. and Olago, D. O. (Eds.) *The East African lakes: Limnology: paleolimnology and biodiversity*, Pp. 62–92, Kluwer, Academic, The Netherlands.

Coppock, D. L., *The Boran plateau of southern Ethiopia: synthesis of pastoral research, development and change, 1980–1991* (Addis Ababa, ILCA, 1993a).

Coppock, D. L., 'Constraints to development of extensive livestock systems: experiences from southern Ethiopia.' *BSAP Occasional Publication* 16 (1993b): 87–92.

Coppock, D. L., Gebru, G. and Desta, S., 'Are cattle die-offs predictable on the Borana plateau?' Research Brief, PARIMA, 2008.

Coppock, D. L., Desta, S., Tezera, S. and Gebru, G., 'Capacity building helps pastoral women transform impoverished communities in Ethiopia,' *Science* 334 (2011): 1394–1397.

Cremaschi, M. and Zerboni, A., 'Human communities in a drying landscape,' in Martini, I. P. and Chesworth, W., *Landscapes and societies* (Dordrecht: Springer, 2010).

Cukierman, D. and Delgrande, J. P., 'Expressing time intervals and repetition within a formalization of calendars,' *Computational Intelligence* 14.4 (1998): 563–596.

Crumley, C.L., 'From garden to globe: Linking time and space with meaning and memory,' in McIntosh et al. (Eds.) The way the wind blows: climate, history, and human action (Columbia University Press, 2000).

Dahl, G., '*Suffering grass: subsistence and society of Waso Borana.*' Stockholm Studies, 1979.

Dahl, Gudrun. 'Women in pastoral production: some theoretical notes on roles and resources,' *Ethnos* 52.1–2 (1987): 246–279.

Dahl, Gudrun, and Anders Hjort. *Having herds: pastoral herd growth and household economy.* Department of Social Anthropology, University of Stockholm, 1976.

Dale, I. R., 'Is East Africa drying up?', *The East African Agricultural Journal*, 17 (3) (1952): 116.

Dapilah, F., Nielsen, J. Ø. and Friis, C., 'The role of social networks in building adaptive capacity and resilience to climate change: A case study from northern Ghana,' *Climate and Development* 12 (1) 2020): 42–56.

Davis, M., *Late Victorian holocausts: El Niño famines and the making of the third world* (London: Verso, 2002, 284).

DeCesare, N. As J., Hebblewhite, M., Bradley, M., Smith, K. G., Hervieux, D. and Neufeld, L., 'Estimating ungulate recruitment and growth rates using age ratios,' *The Journal of Wildlife Management* 76 (1) (2012): 144–153.

Desta, Solomon, and D. Layne Coppock. "Cattle population dynamics in the southern Ethiopian rangelands, 1980–97." *Rangeland Ecology & Management/ Journal of Range Management Archives* 55.5 (2002): 439–451.

Debsu, D. N., 'Customary always in Ethiopia: Institutions of dispute resolution among the Guji Oromo agropastoralists,' *Journal of Oromo Studies Association* 17 (2) (2010): 101–127.

Derbyshire, Samuel F. *Remembering Turkana: Material Histories and Contemporary Livelihoods in North-Western Kenya*. Routledge, 2020.

Diamond, J., Collapse*: how societies choose to fail or survive* (New York and London: Viking Penguin/Allen Lane, 2005).

Donanan H. and Wilson, T. M., 'An anthropology of borders of frontiers,' in Donnan H. and Wilson, T. M., *Border approaches: Anthropological perspectives of frontiers* (University Press of America, 1994a).

Donanan H. and Wilson, T. M., *Border approaches: Anthropological perspectives of frontiers* (University Press of America, 1994b).

Donaldson, T. J. 1986, *Pastoralism, and drought: A case study of the Borana of southern Borana of southern Ethiopia*, MPhil Thesis, University of Reading.

Dong, S., Liu, S. and Wen, L., 'Vulnerability and resilience of human natural systems of pastoralism worldwide,' in Dong, S., Liu, S., and Wen L. (Eds) *Building resilience of human-natural systems of pastoralism in the developing world: Interdisciplinary perspectives*, (Springer International Publishing Switzerland, 2016a).

Dong, S., Kassam, K.-A., Tourand, J. F. and Boone, R. B. (Eds.) *Building resilience of human-natural systems of pastoralism in the developing world* (Interdisciplinary perspectives Springer International Publishing, Switzerland, 2016b)

Dong, S., Liu, S. and Wen, L., 'Vulnerability and resilience of human-natural systems of pastoralism worldwide' in Dong, Shikui, Lu Wen, Shiliang Liu, Xiangfeng Zhang, James P. Lassoie, Shaoliang Yi, Xiaoyan Li, Jinpeng Li, and Yuanyuan Li. 'Vulnerability of worldwide pastoralism to global changes and interdisciplinary strategies for sustainable pastoralism.' *Ecology and Society* 16, no. 2 (2011).

Doran, M. H., Low, A. R., and Kemp, R. L., Cattle as a store of wealth in Swaziland: Implications for livestock development and overgrazing in Eastern and Southern Africa,' *American Journal of Agricultural Economics* 61 (1) (1979): 41–47.

Downing, T. E., Ringius, L., Hulme, M. and Waughray, D., 'Adapting to climate change in Africa,' *Mitigation and Adaptation Strategies for Global Change* 2 (1997): 19–44.

Drake-Brockman, R.E., *British Somaliland* (London: Hurst & Blackett, Ltd., 1912).

Drysdale, J., *The Somali dispute* (New York: Fredrick A. Praeger, Inc., 1964).

Dynes, Russel, *Community social capital as the primary basis for resilience.* Preliminary Paper # 344.

Dynes, R. L. R., 'Community social capital as the primary basis for resilience,' (2005).

Ehret, Christopher. 'Historical Linguistics.' *The International Encyclopedia of Anthropology* (2018): 1–9.

Ellis, J. E., 'Climate variability and complex ecosystem dynamics: Implications for pastoral development,' in Scoones, I. (Ed.) *Living with uncertainty: new directions in pastoral development in Africa* (1994).

Ellis, J. E., and Swift, D. M., 'Stability of African pastoral ecosystems: alternate paradigms and implications for development,' *Rangeland Ecology & Management/Journal of Range Management Archives* 41.6 (1988): 450–459.

Eneyew, A., 'Is settling pastoralists a viable livelihood strategy? Implication for policy dialogue,' *Scholarly Journal of Agricultural Science* 2 (5) (2012): 94–102.

Endfield, H. H., 2008, *Climate and society in colonial Mexico: A study in vulnerability*, Blackwell Publishing, RGS-IBG, Oxford.

Endris, G. S., Kibwika, P., Obaa, B. B., and Hassan, J. Y., 'Does the implementation of social safety net intervention affect indigenous social capital systems for coping with livelihood shocks? Ethnographic evidence of agro-pastoral communities in Eastern Ethiopia,' in Efobi, U. T. and Asongu, S. (eds.) *Financing sustainable development in Africa*, 2018, https://doi.org/10.1007/978-3-3-319-78843-2_11.

Ensminger, J., 'Reputations, trusts, and principal agent problem,' in Cook, Karren, S. (ed.) *Trust in society* (New York, Russel Sage Foundation, 2001).

Esenu, B., Ossiya, S., Oluka, J., Aliu, D., Owesigire, B., Ebiyau, G. and Olokojo, B., 'Positioning agro-pastoral women in livestock production: The link between socio-economic factors and improved household food security,' TROPICULTURA, 10th Anniversary of VSF-Belgium 2005.

Estes, R. J., '"Poverties" and wealth: Competing definitions and alternative approaches to measurement,' in Glatzer, W. (Ed.) *Rich and poor* (Kluwer Academic Publishers, 2002).

Fagan, Brian. *The great warming: Climate change and the rise and fall of civilizations.* Bloomsbury Publishing USA, 2008.

Flores, D., 'Nature's children: Environmental history as human natural history,' in Herron J. P. and Kirk, A. G. (Eds.) *Human/ nature biology, culture and environmental history* (Albuquerque: University of New Mexico Press, 1999).

Forest, R. and Kearns, A., 'Social cohesion, social capital and the neighborhood,' *Urban Studies* 38.12 (2001): 2125–2143.

Fratkin, E., 'East African pastoralism in transition: Maasai, Boran, and Rendille cases,' *African Studies Review* 44 (3) (2001): 1–25.

Fratkin, E., *Ariaal pastoralists of Kenya: surviving drought and development in Africa's arid lands* (Ally and Bacon, Boston USA, 1998),

Fratkin, E. and Roth, E. A., 'Drought and economic differentiation among Ariaal pastoralists of Kenya,' *Human Ecology* 18 (4) (1990): 385–402.

Fratkin, E. and Smith, K., 'Women's changing economic roles with pastoral sedentarization: Varying strategies in alternate Rendille communities,' in *As pastoralists settler*, Springer, Boston, MA, 2005.

Fratkin, E., 'Stability and resilience in East African pastoralism: The Rendille and the Ariaal of Northern Kenya' *Human Ecology* 14 (3) (1986): 264–286.

Galaty, J., 'The Maasai group ranch: politics and development in an African pastoral society,' in Salzman, P. C. (Ed) Galaty, John G., et al. "The future of pastoral peoples." *Conference on the Future of Pastoral Peoples,* Nairobi (Kenya), 4–8 Aug 1980. IDRC, 1981.

Galvin, K. A., Thornton, P. K., Boore, R. B. and Sunderland, J., 'Climate variability and its impacts on east African livestock herders: The Maasai of Ngorongoro conservation area, Tanzania,' *African Journal of Range and Forage Science* 21 (3) (2004): 183–189.

Gies, L., Agusdinata, D. B. and Merwade, V., 'Drought adaptation policy development and assessment in East Africa using hydrologic and system dynamic modelling,' *Natural Hazards* 74 (2014): 789–813.

Gifford, Gonzalez, 'Early pastoralists in East Africa: Ecological and social dimensions,' *Journal of Anthropological Archeology* 17 (1998): 166–200.

Gilliland, H. B., 'An approach to the problem of the government of nomadic peoples: With special reference to experience in Eastern British Somaliland,' *South* Africa *African Geography* 29 (1) (1947): 43–58.

Gitz, V. and Meybeck, A., 'Risks, vulnerabilities and resilience in a context of climate change,' in Meybeck, A., Lankoski, J., Redfern, S., Azzu, N. and Gitz, V. (eds) *Building resilience for adaptation to climate change in the agricultural sector*, Proceedings of a Joint FAO/OECD Workshop, Rome, 23–24 April 2012.

Goldman, M. J., and Riosmena, F., 'Adaptive capacity in Tanzanian Maasailand: Changing strategies to cope with drought in fragmented landscapes,' *Global Environmental Change* 23 (2013): 597.

Graham, O., *Enclosure of the East African rangelands: Contemporary trends and their impact*, ODI, Pastoral Development Network, 1988

Grandin, B. E., 'The importance of wealth effects on pastoral production: a rapid method for wealth ranking,' In: *Pastoral systems research in sub-saharan Africa* (Addis Ababa: ILCA, 1983a), pp. 237–256.

Grandin, B. E., 'The importance of wealth effects on pastoral production: A rapid method for wealth ranking.' In *pastoral systems research in sub-Saharan Africa. The proceedings of workshop held at ILCA*, Addis Ababa, Ethiopia 21 to 24 March 1983b, Pp. 237–254.

Gufu, B. S., 'Sustainable pastoral livelihoods at a crossroads: Broken system resilience and institutional disenfranchisement among the Borana Oromo of southern Ethiopia,' in Muchie, M., Gumede, V., Oloruntoba, S., and Check, N. A. (eds.) *Regenerating Africa: Bringing African solutions to African problems*, Africa Institute of South Africa, 2017.

Gwynn, C. W., 'A journey in southern Abyssinia,' *The Geographical Journal* XXXVIII (1911): 113–134.

Håkanssen, N. T., 'Cattle, climate and caravans: The dynamics of pastoralism, trade, and migration in 19th-century East Africa,' in *Migration in Africa* (London, Routledge, 2022, pp. 95–111).

Handmer, J. W., Dovers, S. and Downing, T. E., 'Societal vulnerability to climate change and variability, mitigation and adaptation,' *Strategies for Global Change* 4 (1999): 267–281

Hardin, R., 'Gaming trust,' in Ostrom and Walker (Eds.) *Trust and reciprocity: Interdisciplinary lessons for experimental research.* Russell Sage Foundation, 2003.

Hardin, G., 'The tragedy of the commons: the population problem has no technical solution; it requires a fundamental extension in morality,' *Science* 162.3859 (1968): 1243–1248.

Hartwig, G. W., 'Oral data and its historical function in East Africa,' *The International Journal of African Historical Studies* 7 (3) (1974): 468–479.

Hartwig, G. W., 'Demographic considerations in East Africa during the late nineteenth century,' *International Journal of African Historical Studies* 4 (1979): 653–672.

Hastenrath, S., 'Variations of East African climate during the past two centuries,' *Climate Change* 50 (2001): 209–217

Heffernan, C., 'Livestock, Destitution and Drought: The impact of restocking on food security post-disaster,' Unpublished (1998).

Helland J., 'Participation and governance in development of Borana: southern Ethiopia,' in Salih, M. A. (Eds). *African pastoralism: Conflict, institutions, and government* 2001.

Helland, Johan. *Five essays on the study of pastoralists and the development of pastoralism,* 20 (1980).

Helland, J. The role of sharing in Borana pastoralism. Paper presented at a workshop on 'Perspectives in survival strategies in African arid lands,' Organized by the Scandinavian Institute of African Studies, Aarhus, Denmark, 3–6 October 1991.

Henige, David Patrick. *The chronology of oral tradition* (Oxford: Clarendon Press, 1974).

Herring, R. S. 'Hydrology, and chronology: the Rodah nilometer as an aid in dating interlacustrine history,' in: F.B. Webster (Ed.), *Chronology, Migration and Drought in Interlacustrine Africa*, London, 1979

Hazell, Peter BR, and Ulrich Hess. 'Drought insurance for agricultural development and food security in dryland areas' *Food Security* 2.4 (2010): 395–405.

Hobson, B. and Lindholm, M., 'Collective identities, women's power resources, and the making of welfare states,' *Theory and Society* 26 (1997): 475–508.

Hogg, R., *Re-stocking pastoralists in Kenya: A strategy for relief and rehabilitation*, ODI, PDN 19C, 1985.

Hoeggel, F. U. and Mbeyale, G., *Impact assessment of pastoralist field schools in Ethiopia and Uganda*, FAO, SADC, 2014.

Hjort, A., *Savanna town: Rural ties and urban opportunities in northern Kenya* (Stockholm, Studies in social anthropology, 1979).

Holden, S. J., Coppock, D. L. and Mulugeta Assefa, 'Pastoral dairy marketing and household wealth interactions and their implications for calves and humans in Ethiopia'. *Human Ecology* 19 (1991): 35–59.

Holling, C. S., 'Resilience and stability of ecological systems,' *Annual Review of Ecology and Systematics* 4 (1973): 1–23.

Holling C.S., Gunderson, L. H. and Ludwig, D., 'Inquest of a theory of adaptive change,' in Gunderson, Lance H., Craig Reece Allen, and Crawford S. Holling, eds. *Foundations of ecological resilience.* Island Press, 2012.

Homewood, Katherine, Trench, Pippa Chenevix and Patti Kristjanson, P., 'Staying Maasai? Pastoral livelihoods, diversification and the role of wildlife in development,' *Staying Maasai?* Springer, New York, NY, 2009a. 369–408.

Homewood, K. and Lewis, J., 'Impact of drought on pastoral livestock in Baringo, Kenya, 1983–85,' *Journal of Applied Ecology* 24 (2) (1987): 615–631.

Homewood, Katherine, Kristjanson, Patti and Trench, Pippa (eds). *Staying Maasai? Livelihoods, conservation, and development in East African rangelands* (Vol. 5. Springer Science & Business Media, 2009b).

Houessou, S. O., Dossa, L. H., Diogo, R. V.C., Houinato, M., Buerkert, A., and Schlecht, E., 'Change and continuity in traditional cattle farming systems of West African coast countries: A case study from Benin,' *Agricultural Systems* 168 (2019): 112–122.

Hulme, M., Doherty, R., Ngara, T., New, M. and Lister, D., 2001. 'African climate change: 1900–2100,' *Climate Research* 17: 145–168.

Huxley, E., *Whiteman's country. Lord Delamere and the making of Kenya, Vol 1 1870–1914* (London: Chatto & Windus, 1953).

Ichani, X. F., Matheka, R. M. and Wario, H. A., 'Debating on success of the Kenya counter insurgency on the shifta war: Between double success and little consolation,' *International Affairs and Global Strategy*, 71 (2019).

Jamshed, A., Birkmann, J., Feldmeyer, D. and Rana, I. A., 'A conceptual framework to understand the dynamics of rural-urban linkages for rural vulnerability,' *Sustainability* 12 (2020): 2894, https://doi.org/10.33010/su/2072894.

Jansen, J., 'Habitation and warfare strategies in 19th century Mande—a view from the Kafu,' in Howard, A. M and Shain, R. M. (Eds.) *The spatial factor in African history: The relationship of the social, material and perceptual* (Leiden: Brill, 2005).

Jost, Christine C., Serge Nzietchueng, Simon Kihu, Bernard Bett, George Njogu, Emmanuel S. Swai, and Jeffrey C. Mariner. 'Epidemiological assessment of the Rift Valley fever outbreak in Kenya and Tanzania in 2006 and 2007,' *The American Journal of Tropical Medicine and Hygiene* 83 (2) (2010): 65.

Lamb, H. H., *Climate, history and the modern world* (London: Routledge, 1982).

Lamphear, J., 'Aspects of becoming Turkana: Interactions and assimilation between Maa and Ateker speakers,' In Spear, T. and Waller, R. (Eds.) *Being Maasai: ethnicity and identity in East Africa* (Ohio University Press, 1993).

Lane, P., 'Trajectories of pastoralism in northern and central Kenya: An overview of the archeological and environmental evidence,' in Bollig, M., Schnegg, M and Wotzka, Ha-P. (Eds.) *Pastoralism in Africa: Past, present and future* (New York: Berghahn, 2013).

Laquinta, D. L. and Drescher, A. W., 'Defining peri-urban: Rural-urban linkages and institutional connections,' ResearchGate Nebraska Wesleyan University, 2000. http://www.researchgate.net/publication/287613842.

Leaning, F. and Guha-Spir, D., 'Natural disasters, armed conflicts and public health,' *The New England Journal of Medicine* 369 (2013): 1836–42.

Legesse, A. *Gada: three approaches to the study of African Society* (New York: The Free Press, 1973).

Le Roy Ladurie, E., *Times of feast, times of famine: A history of climate since the year 1000* (New York: Garden City, 1971).

Lesnoff, M., Cornianx, C. and Hiernaux, P., 'Sensitivity analysis of the recovery dynamics of a cattle population following drought in the Sahel region,' *Ecological Modelling* 232 (2012): 28–39.

Lesorogol, Carolyn K. *Contesting the commons: Privatizing pastoral lands in Kenya*. University of Michigan Press, 2008.

Lesser, E.L. 2000, 'Leveraging social capital in organizations,' In Lesser, E.L., and Wobarn, M.A. (eds.) *Knowledge and social capital: Foundations and applications* (Butterworth-Heinmann, 2000).

Lin, Nan. 'A network theory of social capital,' *The handbook of social capital* 50.1 (2008): 69, 12.

Lind, J. and Barrero, L. R., n.d. *Into the fold: What pastoral responses to crisis tell us about the future of pastoralism in the Horn*, Working Paper 091, Future Agricultures, www.future-agricultures.org.

Laužikas, M., and Dailydaite, S., 'Impacts of social capital on transformation from efficiency to innovation-driven business,' *Journal of Business Economics and Management* 16.1 (2015): 37–51

Ley Roy Ladurie, E., *Times of feast, times of famine: A history of climate since the year 1000* (New York: Garden City, 1971).

Little, M. A., Dyson-Hudson, R., Leslie, P. W. and Dyson-Hudson, N., 'Framework and theory' in Little, M. A., and Leslie, P. W. (Eds) *Turkana herders of the dry savanna: Ecology and biobehavioral response to nomads to an uncertain environment* (Oxford: Oxford University Press, 1999).

Little, P. D., Mahmoud, H. and Coppock, D. L., 'When deserts flood: Risk management and climatic processes among East African pastoralists'.

Little, Peter D., 'Social differentiation and pastoralist sedentarization in northern Kenya, *Africa* 55.3 (1985): 243–261.

Little, P. D., McPeak, J., Barrett, C. B. and Kristjanson, P., 'Challenging orthodoxies: Understanding poverty in pastoral areas of East Africa,' *Development and Change* 39 (4) (2008): 587–611.

Little, P. D., Mahmoud, H. and Coppock, D. L., 'When deserts flood: Risk management and processes among East African pastoralists,' *Climate Research* 19 (2001a): 149–159.

Little P.D., Smith, K., Cellarius, B. A., Coppock, D. L. and Barrett, B., 'Avoiding disasters: Diversification and risk management among East African herders,' *Development and Change* 32 (2001b): 401–433.

Little, P. D., Stone, M. P., Mogues, T., Castro, A. P., & Negatu, W., 'Moving in place': drought and poverty dynamics in South Wollo, Ethiopia,' *The Journal of Development Studies*, *42*(2) (2016), 200–225.

Lunde, T. M. and Lindtjørn, B., 'Cattle and climate in Africa: How climate variability has influenced national cattle holdings from 1961–2008,' *PeerJ*. https://doi.org/10.7717/peerJ-55.

Lybbert, Travis J., Christopher B. Barrett, Solomon Desta, and D. Layne Coppock. 'Stochastic wealth dynamics and risk management among a poor population,' *The Economic Journal* 114, no. 498 (2004): 750–777.

Kamara, A. B., Swallow, B. and Kirk, M. 2004, 'Policies, interventions and institutional change in pastoral resource management of Borana, southern Ethiopia,' *Development Policy Review* 22 (4): 381–403.

Kassahun, A., Snyman, H. A. and Smit, G. N., 'Impact of rangeland degradation on the pastoral production systems, livelihoods and perceptions of the Somali pastoralist in Eastern Ethiopia,' *Journal of Arid Environments* 72 (2008): 1265–1281;

Khalif, Zeinabu Kabale, and Oba, G., "'*Gaafa dhaabaa*-the period of stop': Narrating impacts of shifta insurgency on pastoral economy in northern Kenya, c. 1963 to 2007,' *Pastoralism: Research, Policy and Practice* 3.1 (2013): 1–20.

Khalif, Zeinabu Kabale, and Oba, G., 'Continuity and Change in Social Networks among Borana Women in Northern Kenya,' *Nomadic Peoples* 22.2 (2018): 249–281.

Khazanov, A. 'Pastoral nomads in the past, present and future: A comparative view,' in Olson, P. (Ed.) *The struggles for the land: Indigenous insight and industrial empire in semiarid world* (University of Nebraska Press, 1990).

Kibet, S., Nyangito, M., MacOpiyo, L. and Kefach, D., 'Tracing innovation pathways in the management of natural and social capital on Laikipia Maasai Group ranches, Kenya,' *Pastoralism: Research, Policy and Practice* 6 (2016): 16, https://doi.org/10.1186/s13570-016-006-32.

Kinzig, Ann P., Paul Ryan, Michel Etienne, Helen Allison, Thomas Elmqvist, and Brian H. Walker. "Resilience and regime shifts: assessing cascading effects." *Ecology and Society* 11 (1) (2006).

Koraka, S. and Little, P. D., 'Introduction: Rethinking resilience in the context of East African pastoralism,' *Nomadic Peoples* 25 (2) (2021): 165–180, https://doi.org/10.3197/np.2021.250201.

Kristjanson, P., Neufeldt, H., Gassner, A., Mango, J., Kyazze, F. B., Desta, S., Sayula, G., Thiede, B., Fӧrch, W., Thornton, P. K. and Coe, R., 'Are food insecure smallholder households making changes in their farming practices? Evidence from East Africa,' *Food Science* 4 (2012): 381–397.

Mace, R., Anderson, D. M., Bierschenk, T., Cronk, L., Köhler-Rollefson, I., Lancaster, W., Lancaster, F., Little, P. D., Ann Morris, E. and Rossignol, J., 'Transitions between cultivation and pastoralism in sub-Saharan Africa [and comments and reply],' *Current Anthropology* 34 (4) (1993): 363–382.

Makki, F., 'Power and property: Commercialization, enclosures, and the transformation of agrarian relations in Ethiopia,' *Journal of Peasant Studies* 39 (1) (2021):81–104.

Makki, F., 'Development by dispossession: Terra Nullius and the social-ecology of new enclosures in Ethiopia,' *Rural Sociology* 79 (1) (2014): 79–103.

Mamade, G. F., Womber, B. and Atlew, H., 'The role indigenous social institutions on pastoral and agro-pastoral households' access to and control over livelihood assets: Evidence from Somali region, Ethiopia,' *International Journal of Sciences: Basic and Applied Research*, 30 (5) (2016): 215–225.

Mapiye, C., Chimonyo, M. and Dzama, K., 'Seasonal dynamics, production potential and efficiency of cattle in the sweet and sour communal rangelands in South Africa,' *Journal of Arid Environments* 73 (2009): 529–536.

Markakis, J., *Resource conflict in the Horn of Africa* (London: Sage, 1998).

Martini, I. P., and Chesworth, W., *Landscapes and societies: selected cases.* Dordrecht: Springer, 2010a.

Waller, Richard. 'Ecology, migration, and expansion in East Africa,' *African Affairs* 84.336 (1985): 347–370.

Martyn, K. K. and Belli, R. F., 'Retrospective data collection using event history calendars,' *Nursing Research* 51.4 (2002): 270–274.

Massey, G., 'Subsistence and change,' in Massey, G. (Ed.) *Lessons of Agropastoralism in Somalia.* Boulder: Westview (1987).

Maxwell, S. and Smith, M., *Household food security: A conceptual Review* (Part 1), 1992.

McCabe, J. Terrence. *Cattle bring us to our enemies: Turkana ecology, politics, and raiding in a disequilibrium system* (University of Michigan Press, 2010).

McCabe, J. T. and Ellis, J. E., 'Beating the odds in arid Africa,' *Natural History* 1 (87) (2003): 33–40.

McCabe, J. T., 'Drought and recovery: Livestock dynamics among the Ngisonyoka Turkana of Kenya,' *Human Ecology* 15 (4) (1987): 367–388.

McCann, J. C., *From poverty to famine in northeast Ethiopia: A rural history, 1800–1935* (Philadelphia, PA: University of Pennsylvania Press, 1987, 3).

McPeak, J., Little, P. D. and Doss, C. R., *Risk and social change in an African rural economy: Livelihood in pastoral communities* (London: Routledge, 2012).

McPeak, J. G. and Barrett, C. B., 'Differential risk exposure and stochastic poverty traps among East African pastoralists,' *American Journal of Agricultural Economics* 83 (3) (2001): 674–679.

McPeak, J. G. and Little, P. D., 'Applying the concept of resilience to pastoralist household data,' *Pastoralism: Research, Policy and Practice* 7 (2017): 14.

McIntosh, R. J., Tainter, J. A. and McIntosh, K. 'Climate, history and human action' in McIntosh, R. J., Tainter, J. A. and McIntosh, K. (Eds.) *The way the wind blows: Climate, history, and human action* (New York: Colombia University Press, 2000).

Meikle, S., 'The urban context and poor people,' in Rakodi, C. and Lloyd-Jone, T. (eds.) *Urban livelihoods: A people-centred approach to reducing poverty*, Earthscan Publ. Ltd., London, 2002.

Mengistu, A. 1997 '*The Borana and the 1991–92 drought: A rangeland and livestock resource study*', Institute for sustainable Development, French Catholic Committee against Hunger and for Development, Addis Ababa.

Merryman, J. T., 'The economic impact of war and drought on the Kenyan Somali', *Research in Economic Anthropology* 8 (1987): 249–275).

Middleton, G. D., 'The show must go on: Collapse, resilience, and transformation in 21st-century archeology,' *Reviews in Anthropology* 46.2–3 (2017): 78–105.

Middleton, G. D., 'Nothing lasts forever: Environmental discourses on the collapse of past societies,' *Journal of Archeology Research* 20 (2012): 257–307.

Miller, J. C., 'Introduction: Listening for the African past,' in Miller, J. C. (Ed.) *The African past speaks: Essays on oral tradition and history* (Dawson: Folklore Kent, 1980).

Miller, J. C., 'The significance of drought, disease and famine in the agriculturally marginal zones of west-central Africa,' *Journal of African History* 23 (1982): 17–61.

Miller, D. R., Steele, C. R., Bilgen, I. and Belli, R. F., 'Health status measurement via event history calendars vs. Conventional questionnaires: long time, no status' *AAPOR* (2009): 14–17.

Mohamed, J., 'Epidemics and public health in the late colonial Somaliland,' *Northeast African Studies* 6 (1–2) (1999): 45–81

Monod, Theodore. *Introduction [à Pastoralism in Tropical Africa]*. Oxford University Press, 1975.

Little, Peter D. *The elusive granary: herder, farmer, and state in northern Kenya.* No. 73. Cambridge University Press, 1992.

Moris, Jon R. *Directions in contemporary pastoral development.* Overseas Development Institute, Agricultural Administration Unit, 1986.

Moritz, M., 'Livestock transfers, risk management, and human careers in a West African pastoral system,' *Human Ecology* 41 (2013): 205–219.

Mortzkin, G., Patterson, W. A. and Foster, D. R., 'A historical perspective on pitch pine-scrub oak communities in the Connecticut valley of Massachusetts,' *Ecosystems* 2 (1998): 255–273.

Mulder, M. B., Fazzio, I., Irons, W., McElreath, R. L., Bowles, S., Bell, A., Hertz, T. and Hazzah, L., 'Pastoralism and wealth inequality: Revisiting an old question,' *Current Anthropology* 51 (1) (2010): 35–48.

Mwamfupe, D. G., and Claude G. Mung'ong'o. "Poverty and changing livelihoods of migrant Maasai pastoralists in Morogoro and Kilosa districts." (2003).

Mwangi, Margaret. "Diverse drought spatiotemporal trends, diverse etic-emic perceptions and knowledge: implications for adaptive capacity and resource management for Indigenous Maasai-Pastoralism in the rangelands of Kenya." *Climate* 4.2 (2016): 22.

Namias, J., 1981, 'Severe drought and recent history,' in Rotberg, Robert I., and Theodore K. Rabb (eds.) *Climate and history: studies in interdisciplinary history.* Vol. 341. (Princeton University Press, 2014).

Naess, Johanna Wolf, and Anita Wreford. 'Are there social limits to adaptation to climate change?' *Climatic Change* 93 (3) (2009): 335–354.

Natsagdorji, B. P., and Batnasan, N. 2008, 'Vulnerability of Mongolia's pastoralists to climate extremes and changes,' in Leary, N., Conde, C., Kulkarni J., Nyong, A. and Pulhin, J. (Eds), *Climate change and vulnerability*, Earthscan, London.

Naylor, S., 'Historical geography: Natures, landscapes, environments,' *Progress in Human Geography* 30.6 (2006): 792–802.

Nicholson, M. J. "The cost to productivity and the potential benefits of 2-and 3-day watering of Boran cattle." *Ilca Bulletin* (1986).

Nkedianye, David, et al. "Mobility and livestock mortality in communally used pastoral areas: the impact of the 2005–2006 drought on livestock mortality in Maasailand." *Pastoralism: Research, Policy, and Practice* 1.1 (2011): 1–17.

Nicholson, S. E., 'Climatic and environmental change in Africa during the last two centuries,' *Climate Research* 17 (2001): 123–144.

Ndunda, E. M., Mutinda, M., Murenga, H. and Bor, E., 'Contribution of collective action groups on socio-economic wellbeing of agropastoralists in Makindu sub-County, Kenya,' *International Journal of Contemporary Applied Research* 5 (9) (2018): 199–2019.

Nganga, S. K., Bulte, E. H., Giller, K., Rufino, M.C. and Herero, M., *Exploring use of livestock wealth and social capital by pastoral and agropastoral households in*

ASALs as insurance against climate change and variability risks: A case study of Samburu District in Kenya, Report No. 309-2016-5185, 2013.

Njoki, K. R., Assa, O. C. and Mulama, D., 'Strategies used by women to adapt to the effects of climate change in Mwala Division, Machakos County, Kenya,' in Khalif, Z., Gachene, C. K.K., Gicheru, P., Mburu, D. M. and Gakahu, C. G. (eds.), *Sustainable land management in dry lands of Kenya: Improving land productivity through participatory Research and Technology transfer*, UNDP, Nairobi Kenya, 2015.

Ngoitiko, M., *The pastoral women's council empowerment doe Tanzania's Maasai: The role of local organisations in sustainable development, Gatekeeper* 137e, August 2008.

Nyong, A., Adesina, F. and Elesha, B. O., 'The value of Indigenous knowledge in climate change mitigation and adaptation strategies of African Sahel,' *Mitigation, Adaptation Strategies of Global Change* 12 (2007): 787–797.

Oba, G., *African environmental crisis: A history of science and development* (London: Routledge, 2020).

Oba, G. *Herder warfare in East Africa: A social and spatial history* (Winwick, White Horse Press, 2017).

Oba, G. *Climate change adaptive strategies in Africa: an history ecology* (London, Routledge, 2014).

Oba, G., *Nomads in the shadows of empires: contests, conflicts, and legacies on the southern Ethiopian-northern Kenyan frontier*. Brill, 2013.

Oba, G., 'Mobility and the sustainability of the pastoral production system in Africa: perspectives of contrasting paradigms,' *Future of pastoral peoples in Africa organized by Pastoralism in the Future Agriculture Consortium at ILRI in Addis Ababa, Ethiopia from* (2011): 21–23.

Oba, G. 'The effect of multiple droughts on cattle in Obbu, Northern Kenya.' *Journal of Arid Environments* 49, no. 2 (2001a): 375–386.

Oba, G., 'The importance of pastoralists' indigenous coping strategies for planning drought management in the arid zone of Kenya,' *Nomadic Peoples* 5 (1) (2001b): 89–119.

Oba, G., Vetaas, O. R. and Stenseth N. C., 'Relationships between biomass and plant species richness in arid zone grazing lands,' *Journal of Applied Ecology* 38 (2001): 836–845.

Oba, G., "'Where the bulls fight, it is the grass that suffers': impact of border administration on drought-coping strategies of the Obbu Booran during the 20th century." *Journal of Oromo Studies* 7, no. 1&2 (2000): 87–108.

Oba, G., 'Shifting identities along resource borders: Becoming and continuing to be Borana Oromo,' in Baxter, P.T.W., Hultin J. and Triulizi A. (Eds.) *Being and becoming Oromo: Historical and anthropological enquiries*. Uppsala, Nordiska Africainstitutet, 1996.

Oba, G., 'Changing Property Rights among Settling Pastoralists: An Adaptive Strategy to Declining Pastoral Resources,' in PTW Baxter (ed.) *Property, Poverty, and People: Changing Rights in Property and Problems of Pastoral Development*: 38–44.

Oba, G. *Ecological factors in land use conflicts, land administration and food security in Turkana, Kenya*, London, ODI PN 32a, 1992.

Oba, G. 'Kenya Borana: sharing and surviving, responses of impoverished pastoralists to food insecurity', *Extension Bull. Of the University of Reading* 4 (1994), 17–22.

Obstfeld, D. 'Social networks, the Tertius lungens, orientation, and involvement in innovation,' *Administrative Science Quarterly, 2005* https://doi.org/10.2189/asqu.2005.50.1.100; http://asq.sagepub.com/content/50/100.

Ochieng, R., *Use of seasonal climate forecast and drought effects on livestock assets on Baringo county, Kenya*, PhD Thesis, Egerton University, Nakuru, 2018.

O'Connor, T. G. and Kiker, G. A., 'Collapse of the Mapangwe society: Vulnerability of pastoralism to increasing aridity,' *Climate Change* 66 (2004): 49–66.

O'Leary, M. F., 'Drought and change amongst northern Kenya nomadic pastoralists: The case of the Rendille and Gabra,' in Palsson, G. (Ed.) *From water to world-making* (Uppsala: The Scandinavian Institute of African Studies, 1990a).

O'Leary, M., 'Changing responses to drought in Northern Kenya: The Rendille and the Gabbra.' In Baxter PTW and Hogg, R. (Eds.). *Property, poverty, and people: changing rights in property and problems of pastoral development*` Manchester, UK. Department of Social Anthropology and International Development Centre, University of Manchester, 1990b.

O'Leary, M. F., *The economics of pastoralism in Northern Kenya: The Rendille and the Gabra*. IPAL Technical Report No. F-3. UNESCO-Nairobi, 1985

Olson, J. M., Alagarswamy, G., Andresen, J. A., Campbell, D. J., et al., 'Integrating diverse methods to understand climate-land interactions in East Africa,' *Geoforum* 39 (2008): 898–911.

Oluwafemi, B. A., *Vulnerability assessment of Maasai pastoralists under changing climatic condition*, 2016.

Okello, G. A., and Ngala, O. M., 'Coping strategies among pastoralists women entrepreneurs against the odds: Experiences from Baringo Division, Samburu North, Kenya,' *International Journal of Business and Management* 12 (10) (2017): 166–176.

O'Riordan T. and Lenton, T., *Addressing tipping points for a precarious future*, Published for the British Academy by Oxford University Press, 2013.

Orlove, B., 'The past, the present and some possible futures of adaption,' in Adger, W. Neil, Suraje Dessai, Marisa Goulden, Mike Hulme, Irene Lorenzoni, Donald R. Nelson, Lars Otto.

Ostrom, E., 'Toward a behavioral theory linking trust, reciprocity, and reputation,' in Ostrom and Walker (Eds.) *Trust and reciprocity: Interdisciplinary lessons for experimental research.* Russell Sage Foundation, 2003.

Ostrom, Elinor, and James Walker, eds. *Trust and reciprocity: Interdisciplinary lessons for experimental research* (Russell Sage Foundation, 2003).

Owen-Smith, N., Mason, D. R. and Ogutu, J. O., 'correlates of survival rates for 10 African ungulate populations: Density, rainfall and predation,' *Journal of Animal Ecology* 74 (2005): 774–788.

Owuor, S. O., *Urban households ruralizing their livelihoods: The changing nature of urban-rural linkages in an East African town*, Paper presented at the African Studies Centre Seminar series, Leiden, Thursday 16 December 2004.

Pankhurst, R., *The history of famine and epidemics in Ethiopia prior to the twentieth century* (Addis Ababa: Commission of Relief and Rehabilitation, 1990).

Pannell, D. J., 'Social and economic challenges in the development of complex farming systems,' *Agroforestry Systems* 45 (1999): 393–409.

Parker, D. C., Manson, S. M., Janssen, M. A., Hoffman, J. and Deadman, P., 'Multi-agent systems for the simulation of land use and land cover change: A review,' *Annals of Associations of American Geographers* 93 (2) (2003): 314–337

Pelling, M. and Dill, K., 'Disaster politics: Tipping points for change in the adaptation of sociopolitical regimes,' *Progress in Human Geography* 34 (1) (2010): 21–37.

Pferfferbaum, B., Van Horn, R. L. and Pferfferbaum, R. L., 'A conceptual framework to enhance community resilience using social capital,' *Clin Social J.* 45 (2017), 102–110.

Phiri S. H. 1984, 'National integration, rural development and frontier communities: The case of the Chewa and Ngoni astride Zambian boundaries with Malawi, and Mozambique' in Asiwaju, A. I (Ed.). *Partitioned Africans: ethnic relations across Africa's international boundaries 1884–1984* (Lagos: C. Hurst & Co.).

Pickering, T., *Drought, and Samburu pastoralism: A collaborative effort to examine the causes, adaptation process, and grazing innovations in Northern Kenya*, PhD Dissertation, Colorado State University, Fort Collins, Colorado.

Pratt, D. J. and Gwynne, M. D. 1977 'Rangeland Management and Ecology in East Africa' Hodder and Stoughton, London.

Pretty, Jules, and Hugh Ward. "Social capital and the environment." *World Development* 29.2 (2001): 209–227.

Rabb, T. K., 'The historian and climatologist,' in Rotberg, R. I. and Rabb, T. (Eds.) *Climate and history: Studies in interdisciplinary history* (Princeton University Press, 1981).

Rabb, T. K., 'Climate and Society in History: A Research Agenda,' *Social Science Research and Climate Change* (1983): 61–76.

Rakodi, C., 'A livelihood approach-conceptual issues and definitions,' in Rakodi, C. and Lloyd-Jone, T. (eds.) *Urban livelihoods: A people-centred approach to reducing poverty*, Earthscan Publ. Ltd., London, 2002.

Randall, S., 'Where have all the nomads gone? Fifty years of statistical and demographic invisibilities of African mobile pastoralists,' *Pastoralism: Research, Policy and Practice* (2015): 5: 1–22, https://doi.org/10.1186/s13570-015-0042-9.

Rasmussen, Von L., and Reenburg, A. 'Collapse and recovery in Sahelian agro-pastoral systems: Rethinking trajectories of change'. *Ecology and Society* 17 (1) (2012): 14, https://doi.org/10.5751/ES-04614-170114

Reda, K. T., 'Dynamics in pastoral resource management and conflict in the Borana rangelands of southern Ethiopia,' *African Security Review* 25 (1) :(2016): 31–43.

Redman, C. L. and Kinzig, A., 2003. 'Resilience of past landscapes: Resilience theory, society, and the *longue dure'e*,' *Conservation Ecology* 7.1: 14 [online] http://www.consecol.org/vol17/iss1/art14/.

Reece, G. 'The Horn of Africa,' *International Affairs (Royal Institute of International Affairs 1944–)* 30.4 (1954): 440–449.

Renom, J. G., Mwamidi, D. M. and Domi'nguez, P., 'Holistic ethnographies of East Africa customary pastoral commons needed?' *Current Opinion in Environmental Sustainability* 43 (2020): 83–90.

Rey Craft, R. M. and Bawden, G., 'Environmental disaster and the archaeology of human response: Introduction,' in Bawden, G. and Rey Craft, R. M. (Eds.), *Environmental disaster and the archeology of human response*, Maxwell Museum of Anthropology, Anthropological Papers # 7, Albuquerque, New Mexico.

Richard, Trevor. *Livestock production in central Mali: Long-term studies on cattle and small ruminants in the agropastoral system*. Vol. 14. ILRI (aka ILCA and ILRAD), 1986.

Rigby, P., *Persistent pastoralists: Nomadic societies in transition* (London: Zed Books Ltd., 1985).

Roba, H. (et al.), *Pastoralist mobility in northern Kenya and southern Ethiopia* (Nairobi, 2010).

Robinson, P.W., *Disaster and response among the Gabbra of northern Kenya: An historical perspective*, discussion paper for the conference on ecological stress in East Africa, sponsored by Northwestern University and the National Museums of Kenya, Nairobi, 15–17 June 1980, Unpublished manuscript.

Robinson, C., 'Revisiting the rise and fall of the Somali armed force, 1960–2012,' *Defense and Security Analysis* 32 (3) (2016): 257–252

Robinson, L. W. and Berkes, F., 'Applying resilience thinking questions of policy for pastoral systems: Lessons from the Gabra of Northern Kenya,' *Human Ecology* (2010) 38: 335–350.

Roe, E., Untsinger, L. H. and Labnow, K., 'High reliability pastoralism,' *Journal of Arid Environments* (1998) 39: 39–55

Robert, F. 'The structure of autobiographical memory and the event history calendar: Potential improvements in the quality of retrospective reports in surveys.' *Memory* 6, no. 4 (1998): 383–406.

Rockenbauch, T. and Sakdapolrak, P., 2017. 'Social networks and resilience of rural communities in the Global south: A critical Review and conceptual reflections,' *Ecology & Society* 22 (1): 10. https://doi.org/10.57511/Es-09009-220110.

Rosen, S., Murphy, K. M. and Scheinman, J. A., 'Cattle cycles' *Journal of Political Economy* 102 (3) (1994): 468–492.

Roth, E. A. , 'Traditional pastoral strategies in a modern world: An example from Northern Kenya,' *Human Organization* 55 (1996): 219–224.

Rostila, M., 'The facets of social capital,' *Journal of the Theory of Social Behaviour* 41 (2010): 308–326.

Rubel, P. G., 'Herd composition and structure: On building models of nomadic pastoral societies,' *Man* 4 (2) (1969): 268–273.

Salzman, P.C. *Pastoralists: Equality, hierarchy, and the state* (London: Routledge, 2018).

Salzman, P. C., 'Introduction,' in Shutlive, V. H., Altshuler, N. and Zamora, M. D. (Eds.) *Contemporary nomadic and pastoral peoples: Africa and Latin America* (Williamsburg: Virginia, USA, 1982),

Samuels, M. I., Allsopp, N., and Hoffman, M. T., 'Changes in pastoral mobility in a semi-arid montane region of south Africa,' *African Journal of Range & Forage Science* 38 (1) (2020): 1–13.

Santos, P. and Barrett, C. B., 'Heterogeneous wealth dynamics: On the roles of risk and ability,' in Barrett, C. B., Carter, M. R. and Chavas, J. P. (eds.) *The economics of poverty traps* (Chicago: University of Chicago Press, 2018).

Santos, P. and Barrett, C. 2005. *Herd dynamics, social networks, and informal transfers among southern Ethiopian pastoralists*, Research Brief 05-07, 2005-PARIMA, Global Livestock CRSP.

Schlee, Günther. *Identities on the Move: clanship and pastoralism in northern Kenya* (London, Routledge, 2018).

Schlee, G. and Shongollo, M. 'A Local war and its impact on ethnic and religious identification in southern Ethiopia,' *GeoJournal*, 36 (1995): 7–17.

Scoones, I. and Adwera, A., *Pastoral innovation systems: Perspectives from Ethiopia and Kenya*, Future Agricultures, DFID, FAC Occasional Paper ST101, IDS, UK, 2009

Scoones, I., 'Coping with drought: Responses of herders and livestock in contrasting savanna environments in southern Zimbabwe,' *Human Ecology* 20 (3) (1992): 293–314.

Scott, James C. *The moral economy of the peasant.* Yale University Press (Yale, 1977).

Serneels, S. and Lambin, E. F., 'Proximate causes of land use changes in Narok District, Kenya: A spatial statistical model,' *Agricultural Ecosystems and Environment* 85 (2001): 65–81.

Sellen, D. W., 'Nutritional consequences of wealth differentials in East African pastoralists: The case of the Datoga of Northern Tanzania' *Human Ecology* 31 (4) (2003): 529–565.

Shelter, J. B., *Imaging Serengeti: A history of landscape memory in Tanzania from earliest times to the present* (Ohio University Press, 2007).

Sieff, D. F. 'The effects of wealth on livestock dynamics among the Datoga pastoralists of Tanzania'. *Agricultural Systems* 59, 1999: 1–25.

Sinclair, A. R. E., and J. M. Fryxell. "The Sahel of Africa: ecology of a disaster." *Canadian Journal of Zoology* 63.5 (1985): 987–994.

Schenk, G. J., 'Learning from history'? Chances, problems and limits of learning from historical natural disasters,' in Krűger, F., Bankoff, G., Cannon Terry, Orlowshi and Schipper, L. E. (Eds.) *Cultures and disasters-understanding cultural framing in disaster risk reduction* (London: Routledge, 2015).

Schlee, G., *Identities on the move: clanship and pastoralism in Northern Kenya* (Manchester University Press, 1989).

Schove, D. J., 'African droughts and the spectrum of time,' in Dalby, D., Church, R.J.H. and Bezzaz, F. (Eds.), *Drought in Africa 2*, London, 1977

Schweizer, T. and White, D. R., 'Revitalizing the study of kinship and exchange with network approaches,' in Schweizer, T. and White, D. R. (Eds.), *Kinship, networks, and exchange* (Cambridge University Press, 1998).

Shukla, S., McNally, A., Husak, G. and Funk, C., 'A seasonal agricultural drought forecast system for food-insecure regions of East Africa,' *Hydrology and Earth System Sciences* 18.10 (2014): 3907–3921.

Slavin, P., 'Climate and famines: a historical reassessment,' *WIREs Climate Change* 7 (2016): 433–447.

Slegers, M. F. W., "If only it would rain': Farmers' perceptions of rainfall and drought in semi-arid central Tanzania,' *Journal of Arid Environments* 72 (2008a): 2106–2123.

Smith, A. B., *Pastoralism in Africa: Emergence of pastoral traditions* (London, Hurst & Co., 1992).

Smith, A. D., *Through unknown African countries: The first expedition from Somaliland to Lake Lamu* (New York: Edward Arnold, 1897).

Smith, A. D., 'Dr. Donaldson Smith's expedition in Somaliland,' *The Geographical Journal* 5 (1895): 124–127.

Smit, B. and Wandel, J. 'Adaptation, adaptive capacity and vulnerability,' *Global Environmental Change* 16 (2006): 282–292.

Spencer, Paul. *The Samburu: A study in geocentracy* (London, Routledge, 2004)

Spencer, Paul. *Nomads in alliance: symbiosis and growth among the Rendille and Samburu of Kenya* (Oxford University Press, 2012).

Speranza, Chinwe Ifejika. 'Drought coping and adaptation strategies: Understanding adaptations to climate change in agro-pastoral livestock production in Makueni district, Kenya' *The European Journal of Development Research* 22, no. 5 (2010): 623–642.

Ssemmanda, I., Gelorini, V. and Verschurren, D., 'Sensitivity of East African savannah vegetation to historical moisture-balance variation,' *Climate Past* 10 (2014): 2067–2080.

Shetler, B. J., Imaging *Serengeti: A history of landscape memory in Tanzania from earliest times to the present* (Athens: Ohio University Press, 2007).

Slegers, M.F.W., "If only would it rain": Farmers' perceptions of rainfall and drought in se-arid central Tanzania,' *Journal of Arid Environments* 72 (2008b): 2106–2123.

Shipton, P., 'African famines and food security: Anthropological perspectives,' *Annual Review of Anthropology* 19 (1990): 353–394.

Shivakumar, S. J., 'The place of indigenous institutions in constitutional order,' *Constitutional Political Economy* 14.3 (2003): 3–21.

Spencer, Paul, *The pastoral continuum: The marginalization of tradition in East Africa* (Oxford University Press, 1998).

Spinage, C. A., *African ecology-benchmarks and historical perspectives* (Berlin: Spring-Verlag, 2012, 63).

Soiini, E., *Livelihood, land use and environmental interactions in the highlands of East Africa*, PhD Thesis of University of Helsinki, Finland, 2006.

Stoianovich, T., 'French historical method: The Annales paradigm,' in Stoianovich, T. (Ed.), *French historical method: The Annales paradigm* (Cornell University Press, 1976).

Story, W. T., Tura, H., Rubin, J., Engidawork, B., Ahmed, A., Jundi, F., Iddosa, T. and Abrha, T. H., 'Social capital and disaster preparedness in Oromia, Ethiopia: An evaluation of the "women empowerment" approach,' *Social Science & Medicine* 257 (2020): 111907.

Strayer, D., Glinzenstein, J. S., Jone, C. G., Kolasa, J., Likens, G. E., McDonnell, M. J., Parlar, G. G., and Pickett, S. T. A., *Long term ecological studies: An illustrated account of their design, operation, and importance to ecology*, Occasional Publication of the Institute of Ecosystem Studies (New York: New York Botanical Garden Millbrook, August 1986).

Storås F. (1991) "Cattle complex a begging complex. Livestock transaction and the construction of Turkana society" in Grunhaug, R., Haaland, G. and Henriksen, G. (eds.), *The ecology of choice and symbol Essays in Honour of Fredrick Barth* (Alma Mater Forlag, Norway).

Sutter, D. F. 1987. 'Cattle and inequality: herd size differences and pastoral production among the Fulani of north-eastern Senegal'. *Africa* 57, 198–218.

Sutton, J. E. G. 'Becoming Masai land' in Spear, T. and Waller, R. (Eds.) *Being Maasai: Ethnicity and identity in East Africa* (London: James Currey, 1993).

Martini, I. P. and Chesworth, W., *Landscapes and societies* (Dordrecht: Springer, 2010b).

Swayne, Harald George Carlos. *Seventeen Trips Through Somáliland: A Record of Exploration & Big Game Shooting, 1885–1893: Being the Narrative of Several Journeys in the Hinterland of the Somáli Coast Protectorate, Dating from the Beginning of Its Administration by Great Britain Until the Present Time; with Descriptive Notes on the Wild Fauna of the Country* (London: Rowland Ward, 1895, 188).

Tache, Boku. *Pastoralism under stress: resources, institutions and poverty among the Borana Oromo in southern Ethiopia*. Norwegian University of Life Sciences, Department of International Environment and Development Studies, 2008.

Tache, Boku, and Oba, G., "Is poverty driving Borana herders in southern Ethiopia to crop cultivation?" *Human Ecology* 38, no. 5 (2010): 639–649.

Tache, Boku, and Sjaastad, E., 'Pastoralists' conceptions of poverty: An analysis of traditional and conventional indicators from Borana, Ethiopia,' *World Development* 38.8 (2010): 1168–1178.

Tache, B., *Pastoralism under stress: Resources, institutions and poverty among the Borana Oromo in southern Ethiopia*, PhD Thesis, Department of International Environment and Development Studies, Norwegian University of Life Science Ås, Norway.

Taşan-Kok, Tuna, Dominic Stead, and Peiwen Lu 'Conceptual overview of resilience: History and context,' *Resilience Thinking in Urban Planning* (2013): 39–51.

Temple, R. S. and Rey, I., Livestock population and factors affecting them. In: Nestrel, B. (ed.), *Development of animals' production systems*, world Animal science 2. Elsevier. Amsterdam, 1984.

Tim G., and Gregory Alan Kiker. 'Collapse of the Mapungubwe society: vulnerability of pastoralism to increasing aridity,' *Climatic Change* 66.1 (2004): 49–66.

Tiki, Waktole, and Oba, G. 'Ciinna–the Borana Oromo narration of the 1890s great rinderpest epizootic in Northeastern Africa,' *Journal of Eastern African Studies* 3.3 (2009): 479–508.

Tiki, Waktole, Oba, G. and Terje, T., 'An indigenous time-related framework for reconstructing the impact of disasters on ancient water systems in southern Ethiopia, 1560–1950,' *Journal of Historical Geography* 41 (2013): 33–43.

Tiki, W. and Oba, G., 'Transforming labour and technology of the ancient tula wells for watering livestock in Borana, Ethiopia,' *Nomadic Peoples* 23.2 (2019): 218–240.

Tittonell, P., Maruiki, A., Shepherd, K. D., Mugendi, D., Ksizzi, K. C., Okeyo, J., Verchot L., Coe, R. and Vanlauwe, B., 'The diversity of rural livelihoods and their influence on soil fertility in Agricultural systems of East Africa: A topology of smallholder farms,' *Agricultural Systems* 103 (2010): 83–97.

The Earl of Lytton, *The stolen desert: A study of Uhuru in Northeast Africa* (London: McDonald, 1966).

Thornton, P. K., Galvin, K. A. and Boone, R. B., 'An agro-pastoral household model for the rangelands of East Africa,' *Agricultural Systems* 76 (2003): 601–622.

Tonkin, E., 'Borderline questions: people and space in West Africa,' in Donna H. and Wilson T. M. (Eds.), *Border Approaches: Anthropological Perspectives on Frontiers* (University Press of Amer, 1994).

Toulmin, Camilla. *Livestock losses and post-drought rehabilitation in sub-Saharan Africa* (1985).

Toulmin, C. 1995, 'Tracking through drought: options for destocking and restocking' in Scoones, I. (Eds.) *Living with uncertainty: new directions in pastoral development in Africa* (London: ITP).

Tuffa, S. and Treydte, A. C., 'Modeling Boran cattle populations under climate change and varying capacity,' *Ecological Modelling* 352 (2017): 113–127.

Tronvoll, K., 'Borders of violence-boundaries of identity: Demarcating the Eritrean nation-state,' *Ethnic and Racial Studies* 22 (6) (1999): 1037–1060.

Turner, M. D. and Hiernaux, P., 'Changing access to labor pastures, and knowledge: The extensification of grazing management in Sudano-Sahelian West Africa,' *Human Ecology* 36 (2008): 59–80.

Tynon, Joanne F., Sundstrom, S., and Western, D., 'Rangeland Privatization and the Maasai Experience: Social Capital and the Implications for Traditional Resource Management in Southern Kenya' (2012).

Vansina J. 1980, 'Memory and oral tradition,' in Miller, J. C. (Eds.) *The African past speaks: Essays on oral traditional and history* (Hamden: Archon, 1980).

Van Zwanenberg, R. M. A. and King, A., *An economic history of Kenya and Uganda 1800–1970* (London: Macmillan, 1975).

Verma R. *Gender, land, and livelihoods in East Africa: Through farmer's eyes*, International Development Research Centre, 2001.

Vernon, R. Fox. 'A brief history of resilience,' *Community planning to foster resilience in children*. Springer, Boston, MA, 2004. 13–26.

Osterhoudt, Sarah. 'Remembered resilience: oral history narratives and community resilience in agroforestry systems,' *Renewable Agriculture and Food Systems* 33.3 (2018): 252–255.

Verschusen, D., 'Lake-based climate reconstruction in Africa,' *Hydrobiologia* 500 (2003): 315–330.

Vogel, B. and Henstra, D., 'Studying local climate adaption: Heuristic research framework for comparative policy analysis,' *Global Environmental Change* 31 (2015): 110–120.

Wagenaar, K. T. *Productivity of transhumant Fulani cattle in the inner Niger delta of Mali*. Vol. 13. ILRI (aka ILCA and ILRAD), 1986.

Waller R., '"Emutai": Crisis and response in Maasailand 1883–1902,' in Johnson, D. and Anderson, D. M. (Eds.) *The ecology of survival: Case studies from north-eastern Africa* (Boulder, CO: Westview Press, 1988).

Waters-Bayer, A., 'A local innovation and participatory innovation development: The concepts and their application in livestock-related research: lessons for livestock development,' Proceedings of the 12th Annual Conference of ESAP held in Addis Ababa, August 12–14, 2005.

Watete, P.W., Makau, W. K., Njoka, J. T., Opiyo, L. A. and Mureithi, S. M., 'Are there options outside livestock economy? Diversification among households of northern Kenya,' *Pastoralism: Research, Policy and Practice* (2016) 6: 3, https://doi.org/10.1186/s13570-016-00504.

Watts, M., *Silent violence: Food, famine, and peasantry in northern Nigeria* (Berkeley: University of California Press, 1983).

Webb, J. L. A., *Desert frontier: Ecological and economic change along the western Sahel 1600–1850* (Madison, WI: University of Wisconsin Press, 1995).

Webster, J. B., *Chronology, migration and drought in interlacustrine Arica* (London: Longman University Press, 1979).

Weiss, H., '"Dying cattle": Some remarks on the impacts of cattle epizootics in the central Sudan during the nineteenth century' *African Economic History* 26 (1998): 173–1999.

Wells, M., 'Resilience in older adults living in rural, suburban, and urban areas,' *Journal of Rural Nursing and Health Care* 10 (2) (2010): 45–54.

Western D. and Frinch, V. 1986, 'Cattle and pastoralism survival and production in arid lands,' *Human Ecology* 14: 77–94.

Wilding, R., *The history of pastoralism and the emergence of the Borana Oromo: A review of issues*, Joint Ethiopian Pastoral Systems Study Research Report No. 15 (Addis Ababa: ILCA, 1985).

Whittaker, H., 2008. 'Pursuing pastoralists: The stigma of shifta during the 'shifta war' in Kenya, 1963–68, ERAS Edition 10, http://www.arts.brash.edu.au/publication/eras.

Wood, C.A., A preliminary chronology of Ethiopian droughts, in: D. Dalby, R.J.H. Church and F. Bezzaz (Eds), Drought in Africa, London, 1977.

Woolcock, M., 'The place of social capital in understanding social and economic outcomes,' *Canadian Journal of Policy Research* 2 (1) (2001): 11–17.

Wright D. K., 'Long-term dynamics of pastoral ecology in Northern Kenya: An old model for new resilience,' *Journal of Anthropology Archeology* 55 (2019): 101068.

Wright, Katy. *Community Resilience: A Critical Approach*. Routledge, 2021.

Yosef, T., Mengistu, U., Solomon, A., Mohammed, Y. K. and Kefelgn, K., 'Camel and cattle population dynamics and livelihood diversification as a response to climate change in pastoral areas of Ethiopia' Livestock. *Research for Rural Development* 25 (9) (2013): 1–13.

Xiaogang, S., 'Dynamics of continuity and change in pastoral subsistence among the Rendille in Northern Kenya: With special reference to livestock management and responses to Socio-economic change,' *African Study Monographs* 31 (2005): 1–94.

Zamani Gh. H., Gorgievski-Oujvesteijn, Zafshani K., 2006, 'Coping with drought: Towards a multi-level understanding based on conservation of resource theory,' *Human Ecology*. https://doi.org/10.1007/s10745-006-90340-0.

Zampalingre, N., Dossa, L. H. and Schlecht, E., 'Climate change and variability: Perception and adaptation strategies of pastoralists and agro-pastoralists across different zones of Burkina Faso,' *Reg. Environmental Change* 14 (2014): 769–783.

Index[1]

[1] Note: Page numbers followed by 'n' refer to notes.

G. Oba, *Pastoralist Resilience to Environmental Collapse in East Africa since 1500*, https://doi.org/10.1007/978-3-031-48291-5

www.ingramcontent.com/pod-product-compliance
Ingram Content Group UK Ltd.
Pitfield, Milton Keynes, MK11 3LW, UK
UKHW020835290926
15470UKWH00004B/1358

* 9 7 8 3 0 3 1 4 8 2 9 0 8 *